Exploring Bentley STAAD.Pro
(CONNECT Edition)
(3rd Edition)

CADCIM Technologies

525 St. Andrews Drive
Schererville, IN 46375, USA
(www.cadcim.com)

Contributing Author
Sham Tickoo
Professor
Purdue University Northwest
Department of Mechanical Engineering Technology
Hammond, Indiana, USA

CADCIM Technologies

Exploring Bentley STAAD.Pro CONNECT Edition (3rd Edition)
Sham Tickoo

CADCIM Technologies
525 St Andrews Drive
Schererville, Indiana 46375, USA
www.cadcim.com

ISBN   978-1-942689-74-4

DEDICATION

To teachers, who make it possible to disseminate knowledge
to enlighten the young and curious minds
of our future generations

To students, who are dedicated to learning new technologies
and making the world a better place to live in

THANKS

To employees of CADCIM Technologies and
Tickoo Institute of Emerging Technologies (TIET)
for their valuable help

Online Training Program Offered by CADCIM Technologies

CADCIM Technologies provides effective and affordable virtual online training on various software packages including Computer Aided Design, Manufacturing and Engineering (CAD/CAM/CAE), computer programming languages, animation, architecture, and GIS. The training is delivered 'live' via Internet at any time, any place, and at any pace to individuals as well as the students of colleges, universities, and CAD/CAM training centers. The main features of this program are:

Training for Students and Companies in a Classroom Setting

Highly experienced instructors and qualified engineers at CADCIM Technologies conduct the classes under the guidance of Prof. Sham Tickoo of Purdue University Northwest, USA. This team has authored several textbooks that are rated "one of the best" in their categories and are used in various colleges, universities, and training centers in North America, Europe, and in other parts of the world.

Training for Individuals

CADCIM Technologies with its cost effective and time saving initiative strives to deliver the training in the comfort of your home or work place, thereby relieving you from the hassles of traveling to training centers.

Training Offered on Software Packages

CADCIM provides basic and advanced training on the following software packages:

CAD/CAM/CAE: CATIA, Pro/ENGINEER Wildfire, Creo Parametric, Creo Direct, SolidWorks, Autodesk Inventor, Solid Edge, NX, AutoCAD, AutoCAD LT, AutoCAD Plant 3D, Customizing AutoCAD, EdgeCAM, and ANSYS

Architecture and GIS: Autodesk Revit (Architecture/Structure/MEP), AutoCAD Civil 3D, AutoCAD Map 3D, Navisworks, Primavera, and Bentley STAAD.Pro, ArcGIS, ETABS, MX Road, AutoCAD Raster Design, MS Project

Animation and Styling: Autodesk 3ds Max, Autodesk 3ds Max Design, Autodesk Maya, Autodesk Alias, The Foundry NukeX, and MAXON CINEMA 4D

Computer Programming: C++, VB.NET, Oracle, AJAX, and Java

*For more information, please visit the following link: **http://www.cadcim.com***

Note
*If you are a faculty member, you can register by clicking on the following link to access the teaching resources: **http://www.cadcim.com/Registration.aspx**. The student resources are available at **http://www.cadcim.com**. We also provide **Live Virtual Online Training** on various software packages. For more information, write us at **sales@cadcim.com**.*

Table of Contents

Chapter 3: Structural Modeling Using Tools

Chapter 4: Defining Material Constants and Section Properties

Chapter 5: Specifications and Supports

Chapter 6: Loads

Chapter 7: Performing Analysis, Viewing Results, and Preparing Report

Conversion Table

Conversion Table-Metric/Imperial			
	Unit	**Multiply By (Factor)**	**To Obtain**
Length	Inch	2.54	Centimeter
	Centimeter	0.393	Inch
	Feet	0.301	Meter
	Meter	3.281	Feet
	Kilometer	0.54	Nautical Mile
	Nautical Mile	1.852	Kilometer
	Feet	0.000304	Kilometer
Weight and Mass	Ounce	28.35	Gram
	Gram	0.0353	Ounce
	Pound	0.453	Kilogram
	Kilogram	2.205	Pounds
	Metric Ton	1.102	Ton
Liquid Measures	Fluid Ounce	0.0296	Liter
	Gallon	3.785	Liter
	Liter	0.264	Gallon
Thrust / Pressure	Pounds Force	4.448	Newton
	Newton	0.225	Pound
	Pound per square inch (psi)	6.895	KiloPascal
Temperature	Kelvin	1	Degree Celsius-273.15
	Degree Celsius	1.8	Degree Fahrenheit +32

Preface

STAAD.Pro CONNECT Edition

STAAD.Pro CONNECT Edition, developed by Bentley Systems, is a powerful software used for structural analysis and design. It has various tools that help in modeling 2D and 3D models. These tools analyze and virtually design any type of structure. This enables the users to automate their tasks, and remove the tedious long procedures involved in the manual methods. STAAD.Pro is an effective tool for structural engineers and construction professionals.

STAAD.Pro has an extremely flexible modeling environment that helps in creating accurate models quickly and accurately. It supports broad ranges of Steel, Concrete, Aluminium, and Timber design codes. It is capable of analyzing any structure for static loads, dynamic response, soil-structure interaction, wind, earthquake, and moving loads. STAAD.Pro supports Bentley Rebar, AutoPipe, RAM Connection, STAAD.Foundation, and other software.

Exploring Bentley STAAD.Pro CONNECT Edition is a comprehensive textbook that has been written to cater to the needs of the students and professionals. The chapters in this textbook are structured in a pedagogical sequence, which makes the learning process very simple and effective for both the novice as well as the advanced users of STAAD.Pro. In this textbook, the author explains in detail the procedure of creating 2D and 3D models, assigning material constants, assigning cross-section properties, assigning supports, defining different loads, performing analysis, viewing results, and preparing report. The chapters in the book are punctuated with tips and notes, wherever necessary, to make the concepts clear, thereby enabling the user to create his own innovative projects.

The highlight of this textbook is that each concept introduced in it is explained with the help of suitable examples to facilitate better understanding. The simple and lucid language used in this textbook makes it a ready reference for both the beginners and the intermediate users.

- **Concepts explained with Examples**
  The author has explained the concepts in detail with examples for better comprehension of the processes involved.

- **Tips and Notes**
  The additional information related to topics is provided to the users in the form of tips and notes.

- **Learning Objectives**
  The first page of every chapter summarizes the topics that are covered in that chapter.

- **Self-Evaluation Test and Review Questions**
  Every chapter ends with Self-Evaluation Test so that the users can assess their knowledge of the chapter. The answers to Self-Evaluation Test are given at the end of the chapter. Also, Review Questions are given at the end of chapters and they can be used by instructors as test questions.

- **Heavily Illustrated Text**
  The text in this book is heavily illustrated with screen capture images.

Symbols Used in the Textbook

 Note
The author has provided additional information related to various topics in the form of notes.

 Tip
The author has provided a lot of information to the users about the topic being discussed in the form of tips.

 New
This symbol indicates that the command or tool being discussed is new in this release.

 Enhanced
This symbol indicates that the command or tool being discussed has been enhanced in this release.

Unit System Followed in the Textbook
In this book, the Metric system has been used as the default unit system.

Formatting Conventions Used in the Textbook
Please refer to the following list for the formatting conventions used in this textbook.

- Names of tools, buttons, options, menu, command, pages, and tabs are written in boldface.

  Example: The **Add Beams** tool, the **OK** button, the **File** menu, the **Modeling** tab, the **General** page, and so on.

- Names of dialog boxes, menus, windows, edit boxes, check boxes, and radio buttons are written in boldface.

  Example: The **Property** dialog box, the **Density** edit box of the **Property** dialog box, and so on.

- Values entered in edit boxes are written in boldface.

  Example: Enter **Buildings** in the **Name** edit box.

- Names of the files are italicized.

  Example: *c03_staad_connect_ex1*

Naming Conventions Used in the Textbook

Tool

If you click on an item in a toolbar and a command is invoked to create/edit an object or perform some action, then that item is termed as tool. For example: **Insert Node** tool and **Translational Repeat** tool, refer to Figure 1.

Figure 1 Different tools in the start screen of STAAD.Pro CONNECT Edition

Button

The item in a dialog box that has a 3d shape is termed as Button. For example, **OK** button, **Cancel** button, **Apply** button, and so on.

Dialog Box

In this textbook, different terms are used to indicate various components of a dialog box, refer to Figure 2.

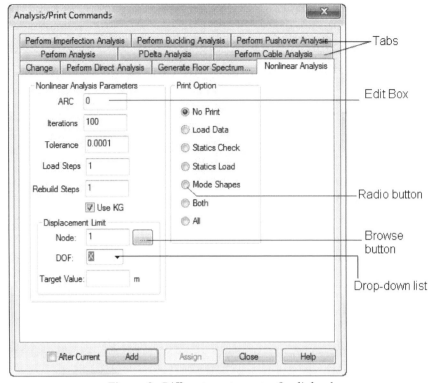

Figure 2 *Different components of a dialog box*

Drop-Down

A drop-down is the one in which a set of common tools are grouped together. You can identify a drop-down with a down arrow on it. These drop-downs are given a name based on the tools grouped in them. For example, **Add Beam** drop-down, **Add Plate** drop-down, and so on; refer to Figure 3.

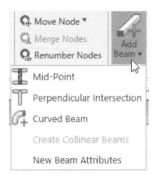

Figure 3 *The **Add Beam** drop-down*

Drop-Down List

A drop-down list is the one in which a set of options are grouped together. You can set various parameters using these options. You can identify a drop-down list with a down arrow on it. To know the name of a drop-down list, move the cursor over it; its name will be displayed as a tool tip. For example, Loading type drop-down list, DDF drop-down list, and so on.

Options

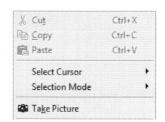

Options are the items that are available in shortcut menus, dialog boxes, drop-down lists, and so on. For example, choose the **Orientation** option from the shortcut menu displayed on right-clicking in the Main Window, refer to Figure 4.

Figure 4 *The shortcut menu displayed in the main window*

Free Companion Website

It has been our constant endeavor to provide you the best textbooks and services at affordable price. In this endeavor, we have come out with a Free Companion website that will facilitate the process of teaching and learning of STAAD.Pro CONNECT Edition. If you purchase this textbook, you will get access to the files on the Companion website.

The resources available for the faculty and students in this website are as follows:

Faculty Resources
• **Technical Support**

   You can get online technical support by contacting *techsupport@cadcim.com*.

• **Instructor Guide**

   Solutions to all review questions and exercises in the textbook are provided in the instructor guide to help the faculty members test the skills of the students.

• **PowerPoint Presentations**

   The contents of the book are arranged in PowerPoint slides that can be used by the faculty for their lectures.

• **Example Files**

   The example files used are available for free download.

Student Resources
• **Technical Support**

You can get online technical support by contacting *techsupport@cadcim.com*.

• **Example Files**

The example files used are available for free download.

If you face any problem in accessing these files, please contact the publisher at ***sales@cadcim.com*** or the author at ***stickoo@pnw.edu*** or ***tickoo525@gmail.com***.

Stay Connected
You can now stay connected with us through Facebook and Twitter to get the latest information about our textbooks, videos, and teaching/learning resources. To stay informed of such updates, follow us on Facebook (***www.facebook.com/cadcim***) and Twitter (***@cadcimtech***). You can also subscribe to our YouTube channel (***www.youtube.com/cadcimtech***) to get the information about our latest video tutorials.

Chapter *1*

Introduction to STAAD.Pro CONNECT Edition

Learning Objectives

After completing this chapter, you will be able to:
* *Understand the basic features of STAAD.Pro CONNECT Edition*
* *Start STAAD.Pro CONNECT Edition*
* *Start new project in STAAD.Pro CONNECT Edition*
* *Explore different components of user interface*
* *Import different files to STAAD.Pro CONNECT Edition*

INTRODUCTION TO STAAD.PRO CONNECT EDITION

STAAD.Pro CONNECT Edition is used to create, analyze, and design any type of virtual structure through its flexible modeling environment. The three basic activities which are to be carried out to achieve this goal are: model generation, calculations to obtain the analytical results, and result verification. All these activities are discussed individually in different chapters. STAAD.Pro is designed for engineers who understand the process of modeling, analyzing, and designing a structure.

BASIC FEATURES OF STAAD.PRO CONNECT EDITION

The basic features of STAAD.Pro CONNECT Edition are listed below:

1) State-of-the art 2D/3D graphical environment.
2) Ability to perform structural analysis and design generate 2D/3D CAD models.
4) Modeling of truss and beam members, plates, solids, linear and non-linear cables, and curvilinear beams.
5) Advanced automatic load generation facilities for wind, area, floor, and moving loads.
6) Customizable structural templates for creating a model.
7) Toggle display of loads, supports, properties, and structural elements.
8) User-controlled numbering scheme for structural elements.

STARTING BENTLEY STAAD.PRO CONNECT EDITION

You can start STAAD.Pro CONNECT Edition by double-clicking on the STAAD.Pro CONNECT Edition icon available on the desktop. Alternatively, choose **Start > Bentley Engineering > STAAD.Pro CONNECT Edition** from the task bar (for Windows 10), refer to Figure 1-1; the start page screen of STAAD.Pro CONNECT Edition will be displayed, as shown in Figure 1-2.

The start screen of STAAD.Pro CONNECT Edition consists of two panes. The left pane contains five tabs: **Open**, **New**, **Help**, **License** and **Configure** and the right pane comprises three areas: **Properties**, **CONNECT Properties** and **Additional Licenses**, refer to Figure 1-2. These tabs and areas are discussed next.

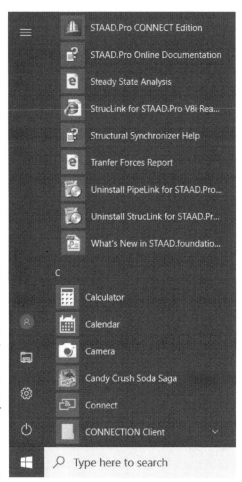

Figure 1-1 *Starting STAAD.Pro CONNECT Edition from the task bar*

Figure 1-2 *The start page of STAAD.Pro CONNECT Edition*

The **Open** tab is used to open an existing STAAD input file. Note that you can also open STAAD projects from a ProjectWise data source. The **Open** tab in the left pane is chosen by default and the **Open** page is displayed in the start screen with four different options: **Recent**, **Projectwise**, **Archive**, and **Open Other Models**. The **Recent** option will be selected by default on choosing the **Open** tab and recently used files will be displayed on the screen. Choose the file either from the recent files or you can also open the model from the archieve files. The **Archive** option is used to create or extract STAAD archive files. Choose the **Open Other Models** option; the **Open** dialog box will be displayed. Browse to the required location and then select the required file from the available list of files. Next, choose the **Open** button; the project will be loaded and structure will be displayed in the **Whole Structure** Window. The **New** tab is used to create a new project. The process of creating a new project is discussed in detail later in this chapter.

The **Help** tab contains several options that enables you to access help documents online or offline, technical support services, product news, and so on. You can access the offline help by clicking on the **Help Contents** option.

The **License** tab is used to check the activation status of currently installed products.

In STAAD.Pro, there are two base unit systems: English Imperial and metric. The default unit system will be the one which you have selected while installing the program. You can also change the base units in the program. To do so, choose the **Configure** option; the **Application Configuration** dialog box will be displayed, refer to Figure 1-3. In this dialog box, the **General** tab and **Base unit** area will be chosen by default. In this page, the current base unit will be displayed at the top. You can change the unit by selecting the **English** or **Metric** radio button from the **Base Unit** area. The **Configure** tab is also used to configure the program settings such as units, color, input or output file formats, default design codes, and so on.

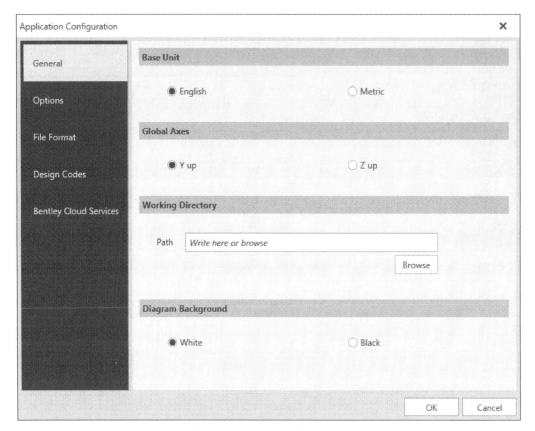

Figure 1-3 *The **Application Configuration** dialog box*

When you hover the cursor on any of the recently used files, its preview and location will be displayed in the **Properties** or **CONNECT Properties** area in the right pane of the user interface. The **Additional Licenses** area consists of three check boxes: **STAAD.Beava (Bridge Codes)**, **STAAD.Planwin**, and **RAM Connection**. You can select the required code before opening a file by selecting the corresponding check box. If the license for the corresponding code is available, the circle besides the check box will be green otherwise will be red.

STARTING A NEW PROJECT

To start a new project, choose the **New** option from the left pane; the **Model Information** page will be displayed, as shown in Figure 1-4. Enter name of the file in the **File Name** edit box in the **Model Information** page. To specify the file location, click on the **Browse** button available on the right of the **Location** edit box; the **Select Folder** dialog box will be displayed. Select the folder where you want to save the file and then choose the **Select Folder** button; the path of the location will be displayed in the **Location** edit box. Next, choose **Analytical** in the **Type** area and **Metric** in the **Units** area.

Figure 1-4 *The start screen of STAAD.Pro CONNECT Edition with the **Model Information** page*

Choose the Job Info option from the New area; the Job Information page will be displayed. In this page, enter the job name, client name, job number, engineer name, project ID, and project name in the respective edit boxes. Choose the **Associate Project** button to save the details before creating the model, refer to Figure 1-5. After specifying all the data, choose the **Create** option in the **New** area; the user interface will be displayed, as shown in Figure 1-6.

Figure 1-5 *The start screen of STAAD.Pro CONNECT Edition with the **Job Information** page*

Figure 1-6 The user interface screen of STAAD.Pro CONNECT Edition

WORKING IN USER INTERFACE

The user interface of STAAD.Pro CONNECT Edition elements are exclusively designed to provide an easy access to the tools and windows. The interface comprises several elements such as the Ribbon, Quick Access Toolbar, Workflows, Data area, View window, Tool Search, Status Bar, and so on, refer to Figure 1-6. Some of these elements are discussed next.

Ribbon Tab

The Ribbon tab is located at the top of the interface. It comprises different tabs: **File**, **Geometry**, **View**, **Specification**, **Loading**, **Analysis and Design**, and **Utilities**. These menus will be available in the **Analytical Modeling** mode. When you switch to a different mode, such as **Postprocessing**, different tabs will appear in the Ribbon tab. These modes are discussed in the later chapters. The different tabs in the Ribbon tab are discussed next.

File

The **File** menu contains the options which are used for performing different file operations such as creating new structure, opening an existing model, saving files, printing files, and so on. Figure 1-7 shows different options of the **File** menu.

The selection of the **Info** option from the **File** menu will display the **Info** page. You can use this page to specify general information about the structure including a job description, job number, persons responsible for creating, checking, and approving the structure, and so on. The **New** option will allow you to create a new STAAD project.

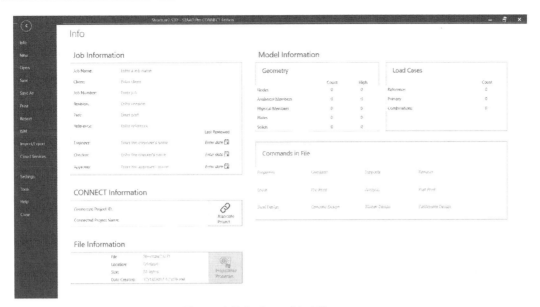

*Figure 1-7 Options of the **File** menu*

The Open option is used to open an existing STAAD input file or archive. The **Save** and **Save As** options are used to save changes to the current model or to save the current model as a different file, respectively. The **Print** option is used to print input, output, and reports for the current STAAD project. The **Report** option is used to configure, review, and export STAAD reports. The **ISM** option contains tools for working with the Integrated Structural Modeling repositories.

The **Import/Export** option will allow you to import data from other formats and also export data to other formats such as DXF, VRML, CIMsteel STEP format, and so on. The **Cloud Services** option opens the Bentley Cloud Services page for using the cloud services. The **Settings** option is used to control some of the application settings.

The **Tools** option displays the **Tools** page which contains tools for managing file backups. The **Help** option displays the **Help** page that has links for getting additional assistance and information about STAAD.Pro. The **Close** option is used to close STAAD graphical environment.

Geometry

The **Geometry** tab contains the Selection, Clipboard, Structure, Node, Beam, Plate, Solid, Physical Member, and Composite deck panels. The tools available in the panels of the **Geometry** tab are shown in the Figure 1-8 and some of these are discussed next.

The **Node Cursor**, **Beam Cursor**, and **Plate Cursor** tools are used to select nodes, beams, and plate element in a structure, respectively.

*Figure 1-8 Options in the **Geometry** tab*

The **Geometry Cursor** tool is used for selecting the nodes, members and elements of the structure at the same time. The **Snap/Grid Node** tool is used to specify the grid and snap settings for creating nodes, beams, plates, and solids. The **Insert Node** tool is used to insert node on an existing member. The **Add Plate** tool is used to add plates by connecting nodes. The **Create Infill Plates** tool is used to automatically generate the floor slab by selecting some or all beams in a structure. The **Translational Repeat** tool is used to copy the entire structure linearly. The **Circular Repeat** tool is used to copy the entire structure in circularly. The **Generate Surface Meshing** tool is used to create a finite element mesh. The **Rotate** tool is used to rotate the selected portions of the structure or the entire structure about the specified axis through a specified distance. The **Merge Selected Members** tool is used to merge two members and replace them with one. The **Renumber** tool is used to renumber nodes, members, or plates starting with a specified number. The **Run Structure Wizard** tool is used to display the **StWizard** window which is discussed in later chapter.

View

The **View** tab consists of the **Labels**, **Tools**, **Views** and **Windows** panels. The **Labels** panel consist of tools which are used to customize the view of the structure by setting different view-related parameters. The **Tools** panel is used to perform various view related tasks such as zooming, panning, setting colors and fonts, and so on. Figure 1-9 shows different options in the **View** tab. The **3D Rendering** option is used to view the rendered view of the structure.

Figure 1-9 Options in the View tab

Specification

The **Specification** tab consists of the **Beam Profiles**, **Plate Profiles**, **Materials**, **Specifications**, **Supports,** and **Tools** panels. The various options in the **Specification** tab are shown in the Figure 1-10.

Figure 1-10 Options in the Specification tab

The **Assign** option is used to assign sections, materials, material properties, supports, and specifications to the structural members. The **Plate Thickness** option is used to specify the plate thickness. The **Constants** option is used to define and assign material constants such as Density, Elasticity, Poisson's Ratio, Coefficient of Thermal Expansion, and so on. The **Section Wizard** option is used to display the **Section Wizard** window. In this window, you can calculate section property values such as area, moment of inertia, and so on for cross sections. These cross sections can be assembled from pre-existing standard shapes, user-created shapes, and parametric shapes.

Loading

The **Loading** tab is used to define and apply loads on a structure. The **Loading** tab consists of the **Loading Specifications**, **Load Generation**, **Define Load Systems**, **Dynamic Specifications**, and **View** panels. The options in these panels are shown in Figure 1-11.

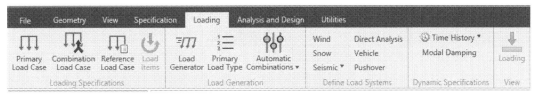

*Figure 1-11 Options in the **Loading** tab*

Analysis and Design

The **Analysis and Design** tab consists of the **Analysis Data**, **Analysis**, and **Design Commands** panels. The various options in these panels are shown in Figure 1-12.

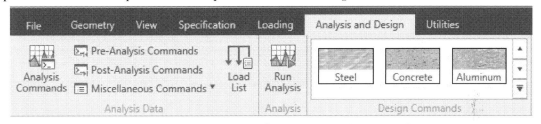

*Figure 1-12 Options in the **Analysis and Design** tab*

The **Analysis Commands** and **Pre Analysis Commands** options open the **Analysis/Print Commands** dialog box which is used to define the analysis and pre-analysis print commands to be included in the input command file, respectively. The **Post Analysis Commands** option opens the **Post Analysis Print -- Whole Structure** dialog box, which is used to define the post-analysis print commands to be included in the input command file. The **Miscellaneous Commands** option is used to add commands such as Input Width, Output Width, Set Z Up, and so on. The **Run Analysis** option performs the STAAD analysis as directed by input commands. The **Design Commands** panel is used to specify design parameters and commands.

Utilities

The **Utilities** tab consists of the **Geometry Tools**, **Physical Model**, **Query**, **Display**, **Utilities**, **Developer**, and **User Tools** panels. The options in these panels are shown in Figure 1-13. The options available in the **Utilities** panel are discussed next.

*Figure 1-13 Options in the **Utilities** tab*

The **Editor** option opens the current input command file (file extension .std) in the **STAAD. Pro CONNECT Edition Editor** window. On selecting the **STAAD Output** option, the results of analysis and design will get displayed in the **STAAD Output Viewer** window. **Unit Converter** opens the **STAAD.Pro Converter** window which is used to convert data from one unit system to another. **AVI File** opens the **Create AVI File** dialog box which is used to create a video file recording for animated deflection, section displacement, mode shape, and plate stress contour diagrams.

Tool Search

You can search for any tool by typing a part of the name in the Search field. In the search results drop-down, click **Show Details** to display the tool tip description for each tool in the results list. The tool search box is placed at the right extreme corner of the Ribbon tab, refer to Figure 1-6.

Quick Access Toolbar

The Quick Access Toolbar is located just above the Ribbon tab. It contains some of the most frequently used tools in a convenient location. Figure 1-14 shows the tools in the Quick Access Toolbar.

Figure 1-14 *Tools in the **Quick Access Toolbar***

 The **Save** tool saves any change made to the current model.

 The **Open** tool opens the start page screen of STAAD.Pro CONNECT Edition with the **Open** tab selected, which is used to select a model to open in the program.

 The **Close** tool closes the current model and returns to the start screen of STAAD.Pro CONNECT Edition.

 The **Undo** tool undo some previous operation.

 The **Redo** tool redo some previous undo operation.

 The **Command File Editor** tool opens the current input command file (file extension .std) in STAAD.Pro CONNECT Edition Editor. If any change made in the structure is unsaved, you are prompted to save the structure first.

 The **STAAD Output** tool opens the results of a successful analysis and design of the current model in the **STAAD Output Viewer** window.

Workflows

The STAAD.Pro user interface allows you to switch between the **Analytical Modeling**, **Physical Modelling**, **Building Planner**, **Piping**, **Bridge Deck**, **Postprocessing**, **Foundation**

Design, **Steel Design**, **Connection Design**, **Concrete Design**, **Advanced Slab Design,** and **Earthquake** tabs, refer to Figure 1-15. Some of these tabs are discussed next.

Figure 1-15 *Partial view of the*
Workflows *tab*

Analytical Modeling

The **Analytical Modeling** tab is the first tab of the **Workflows**, as shown in Figure 1-15. This tab contains different pages which are used for creating structure, assigning properties, assigning supports, defining loads, and so on, refer to Figure 1-16. The pages that are associated with analytical modelling are discussed in detail in the later chapters.

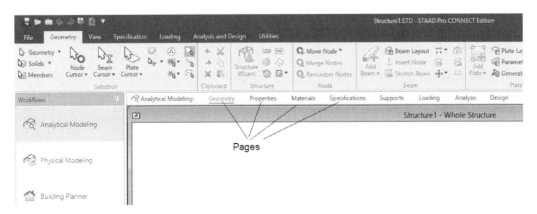

Figure 1- 16 *Pages associated with the analytical modeling*

Physical Modeling

STAAD.Pro Physical Modeler is user interface for modeling structures using physical concepts. Beams and surfaces are placed in the model on the same scale as they would appear in the physical world. You can model structure geometry, specifications, and many loads with the STAAD.Pro Physical Modeler.

Note

*Existing analytical models cannot be opened in the physical modeler. When you open a model that was created using the physical modeler, you will be asked if you want to open the model in the physical modeler environment. Select the **Physical Modeling** workflow. The **STAAD.Pro Physical Modeler** window will be displayed.*

When a structure is to be analyzed, an analytical model is generated from the physical model in a process referred to as decomposition. Surface entities will mesh into plate elements and beams will be automatically subdivided for analysis as required by intersecting members. Physical modeling is discussed in detail in the later chapters. Figure 1-17 shows the user interface screen in physical modeling.

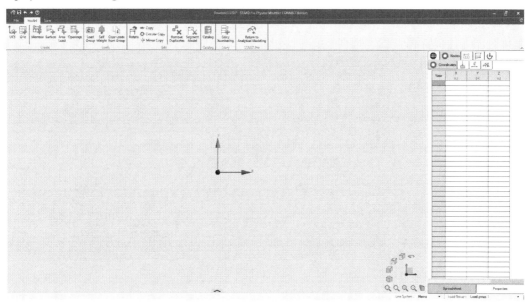

Figure 1-17 User interface screen in Physical modeling

Postprocessing

This workflow offers graphical result verification and visualization facilities. A comprehensive custom report generation facility is also incorporated. The customized reports may be in both tabular and graphic form. The **Postprocessing** tab has the **Displacements**, **Reactions**, **Beam Results**, **Plate Results**, **Solid Results**, *Dynamics*, and **Reports** pages, refer to Figure 1-18. Some of these are discussed next.

*Figure 1-18 Pages in the **Postprocessing** tab*

Displacements

The **Displacement** page allows you to view the nodes and beam relative displacements. You can view them graphically in the view window or in tabular form in the data area.

Beam Results

The **Beam** page allows you to view the member end forces and member stresses in both graphic and tabular form. It also allows you to view the bending moment diagram and shear force diagram of individual members at a time. These pages are discussed in detail in the later chapters.

Reports

The **Reports** page allows you to create customized report. This report will include structural elements, properties, load cases, mode shapes, numerical and graphical results, and so on.

Data Area

In STAAD.Pro, the data area is located on the right side in the interface. This area contains different windows and tables. In this area, you can provide the coordinates for creating nodes and members, define member properties, define supports, define loads, and so on. This area will be displayed on creating a new file or opening an existing file, Figure 1-19 shows partial view of the data area which contains the **Nodes** and **Beams** tables.

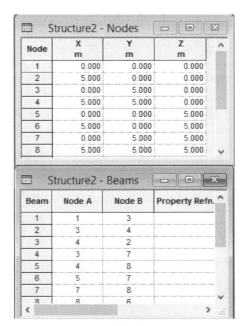

Figure 1-19 Partial view of the **Nodes** and **Beams** tables in the data area

Status Bar

Status Bar displays the current workflow along with tips on next actions based on the current tools. You can also set the current load case and input units in the status bar.

View Window

The View window covers the largest area in the interface. In the View window, the created structure along with the assigned properties, loads, supports, and specifications will be displayed. The analysis results and diagrams will also be displayed in this area.

KEYBOARD SHORTCUTS

In STAAD.Pro, keyboard shortcuts are assigned to some of the frequently used commands. These shortcuts can be typed using the keyboard to invoke the corresponding command. These shortcut keys are also available in the **Labels** tab of the **Diagrams** dialog box, as shown in Figure 1-20. To invoke the **Diagrams** dialog box, right-click in the main window; a shortcut menu will be displayed. Choose the **Labels** option; the **Diagrams** dialog box will be displayed. Table 1-1 shows some of the frequently used shortcut keys in STAAD.Pro.

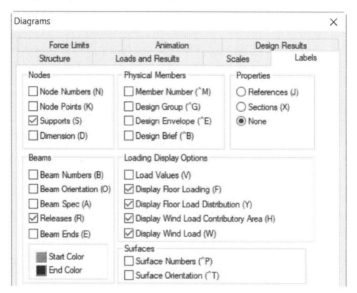

Figure 1-20 *Parial view of the shortcut keys displayed in the* **Labels** *tab of the* **Diagrams** *dialog box*

Table 1-1 *Various shortcut keys used in STAAD.Pro*

Keyboard Shortcut	Description
SHIFT+N	Displays node numbers
SHIFT+B	Displays beam numbers
SHIFT+S	Displays support icons
SHIFT+E	Displays member ends
SHIFT+O	Displays beam orientation
SHIFT+P	Displays plate numbers
SHIFT+L	Displays master slave node

SHIFT+V	Displays load values
SHIFT+C	Displays solid numbers
SHIFT+M	Displays material
SHIFT+I	Displays axes at origin

IMPORTING AND EXPORTING A MODEL IN STAAD.Pro

In STAAD.Pro, data can be imported from the following file types: DXF, QSE ASA, Stardyne, and CIS/2. From the CIS/2 file type, data such as member properties, material constants, member end conditions, support conditions, and loading information can be imported. To import data from any of the files, choose the **Import/Export** option from the **File** menu; the **Import/Export** screen will be displayed, as shown in Figure 1-21.

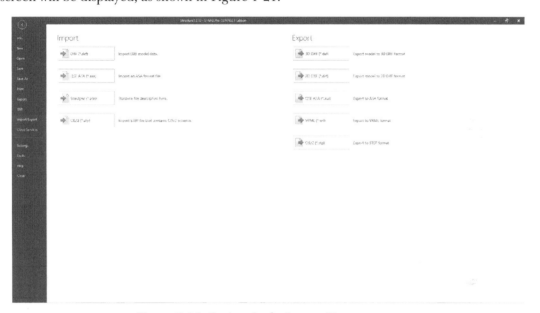

*Figure 1-21 Options in the **Import/Export** screen*

In this screen, select the required file format from the **Import** area by selecting the corresponding radio button. For example, choose the **DXF (*.dxf)** button to import the file from a .dxf file; the **Open** dialog box will be displayed. In this dialog box, browse to the required location, select the file, and choose the **Open** button; the **DXF Import** dialog box will be displayed, as shown in Figure 1-22.

In this dialog box you will define the axis of gravity in your STAAD.Pro model. If the axis system is same in the STAAD.Pro model and the dxf file then choose the **No Change** radio button. Now, choose the **OK** button; the file will be imported to STAAD.Pro.

Figure 1-22 The **DXF Import** *dialog box*

Similarly, you can export the model to the following file types: 3D DXF, 2D DXF, QSE ASA, VRML, and CIS/2. From the 3D DXF file type, only line diagrams can be exported. From the CIS/2 file type, data such as member properties, material constants, member end conditions, support conditions, and loading information can be exported. To export data to any of the files, choose the **Import/Export** option from the **File** menu; the **Import/Export** screen will be displayed, as shown in Figure 1-21. In this screen, select the required file format by choosing the corresponding button. For example, choose the **3D DXF (*.dxf)** button to export the file in a .dxf format; the **Save As** dialog box will be displayed. In this dialog box, browse to the required location, enter the name in the **File name** edit box, and choose the **Save** button; the **DXF Export** dialog box will be displayed, as shown in Figure 1-23. In this dialog box, you will define the axis of gravity in your STAAD.Pro model. If you want the axis system to be same in the STAAD.Pro model and the dxf file, then select the **No Change** radio button and choose the **OK** button; the file will be exported from STAAD.Pro.

Figure 1-23 The **DXF Export** *dialog box*

Self-Evaluation Test

Answer the following questions and compare them to those given at the end of this chapter:

1. Which of the following tabs is used to open an existing STAAD Input file?

    (a) **Configure**   (b) **Open**
    (c) **New**   (d) None of these

2. Which one of the following tools is used for selecting the nodes, members, and elements of the structure at the same time?

    (a) **Node Cursor**   (b) **Beam Cursor**
    (c) **Geometry Cursor**   (d) None of these

3. The _____ shortcut key is used to display the load values.

4. The _____ shortcut key is used to display the member ends.

5. STAAD files are saved in the _____ format.

6. Job information is provided in the **Job Info** dialog box. (T/F)

7. In the **Postprocessing** mode, you can verify the analysis results both graphically and numerically. (T/F)

Review Questions

Answer the following questions:

1. Which of the following shortcut keys is used for displaying the master slave node?

    (a) SHIFT+L   (b) SHIFT+N
    (c) SHIFT+B   (d) SHIFT+V

2. Which of the following commands can be added using the **Miscellaneous Commands** option in the **Analysis and Design** tab?

    (a) Perform Analysis   (b) Run Analysis
    (c) Input Width   (d) Pre Analysis

3. Which of the following panels is not a part of the **Geometry** tab?

    (a) Selection   (b) Structure
    (c) Node   (d) Geometry Tools

4. Postprocessing Mode is used to model the structure. (T/F)

5. Status bar is present in the Ribbon tab. (T/F)

Answers to Self-Evaluation Test
1. c, **2.** b, **3.** SHIFT+V, **4.** SHIFT+E, **5.** .std, **6.** T, **7.** T

Chapter 2

Structural Modeling in STAAD.Pro

Learning Objectives

After completing this chapter, you will be able to:
• *Create structures using Editor*
• *Create structures using STAAD GUI*
• *Create structures using the Structure Wizard*

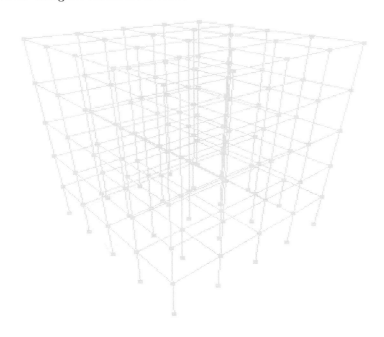

INTRODUCTION

Before erecting a structure, you need to run a stability check for the structure. To check the structural stability, you need to model the structure virtually. The virtual structural models comprise of various components such as beams, columns, walls, slabs, and so on. In STAAD.Pro, you can model a structure and then check its structural stability. A structural model in STAAD.Pro is represented as a line structure which consists of nodes and members. Before modeling a structure, you need to determine the nodes and the connection between them.

In STAAD.Pro, nodes are the joints which are capable to resist forces and moments. A node is located at the end of beams, columns, or plates. A node is always represented by a node number. To create nodes, you need to determine its coordinates in the XYZ space. After creating nodes, you can form members or plate elements by connecting the nodes. A member may be a column, beam, or truss, which is generated by connecting two nodes and is represented by a member number. Figure 2-1 shows the nodes and members in a portal frame structure.

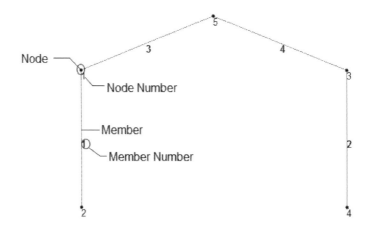

Figure 2-1 *Nodes and members in a portal frame structure*

A plate element is formed by connecting three or four nodes and is represented by a plate number. A plate element can be triangular or quadrilateral. These members and plate elements are represented by member number and element number.

In STAAD.Pro, you can create a structure model by using any of the following methods:

1) Using Editor
2) Using STAAD GUI
3) Using Structure Wizard
4) Building planner

These methods are discussed next.

STRUCTURAL MODELING USING EDITOR

In STAAD.Pro, the **STAAD Editor** window contains a list of commands required to model, analyze, and design a structure. These commands are the instructions related to the analysis and design of a structure and are executed consecutively. To open the **STAAD Editor** window, choose the **Editor** tool from the **Utilities** panel of the **Utilities** tab. Figure 2- 2 shows the STAAD Editor window with default commands entered in it.

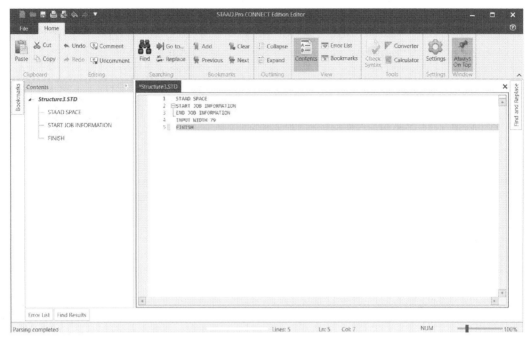

*Figure 2-2 The **STAAD Editor** window*

To start the structure geometry, the nodes must be created first and then other elements can be created like beams, plates, surfaces and solids. The command for creating the nodes is discussed next.

Creating Nodes

Nodes are created by specifying the X, Y, and Z coordinates of the joints in a structure. The command used for specifying the coordinates is **JOINT COORDINATES**. This command will be specified after specifying the unit command. The general format for specifying the coordinates is given below:

```
JOINT COORDINATES
n x y z
```

In the above command, **n** represents the node number and **x**, **y**, and **z** represent the coordinates of the joints nodes in a structure. To create nodes, first you need to specify the command **JOINT COORDINATES** in the **STAAD Editor** window. Next, you will specify the node number and coordinates for the first node. After specifying the first node, you need to put a semi-colon and then specify the node number and coordinates of the second node, and so on. After specifying the command, choose the **Save** button from the toolbar in the **STAAD Editor** window and then close the window. You can view the node numbers of the created nodes by pressing SHIFT+N.

Note
*The **JOINT COORDINATE** command is not case sensitive. You can also write the initial three letters for the command. For example, enter **JOI COO** for this command. In case, you forget to choose the **Save** button before closing the window then the **Save** dialog box will automatically be displayed and you can save the changes by choosing the **Yes** button.*

You can also generate multiple nodes at a time by using the REPEAT and REPEAT ALL commands. These commands generate the nodes in a repetitive pattern.

Note
*The **REPEAT** command repeats the previous line of input n number of times with the increments specified in x, y, and z directions. The **REPEAT ALL** command repeats all the previously specified inputs. When you use **REPEAT** command after **REPEAT** or **REPEAT ALL** then the **REPEAT** command is only applicable for the last created node.*

The general format for specifying the REPEAT and REPEAT ALL commands is given below:

```
JOINT COORDINATES
n x y z
REPEAT n x_i y_i z_i
REPEAT ALL n x_i y_i z_i
```

In the above command, **n** represents the number of times the nodes will be repeated and x_i, y_i, and z_i represent increments in the **x**, **y**, and **z** coordinates.

Note
1. In this chapter, you need to download the c02_staad_connect.zip file for the examples from http://www.cadcim.com. The path of the file is as follows: Textbook > Civil/GIS > STAAD.Pro > Exploring Bentley STAAD.Pro CONNECT Edition.

2. Before starting the examples, you need to create a folder with the name STAAD Examples in C: drive and then extract the downloaded zip folder to this folder.

Example 1

In this example, you will create nodes for a portal frame structure. This structure will be created as a plane frame structure.

Steps required to complete this example are given below:

Step 1: Start STAAD.Pro CONNECT Edition and choose the **New** option from the left pane of the user interface screen; the **Model Information** page is displayed. In this page, specify the name *c02_staad_connect_ex1* in the **File Name** edit box and browse to the location *C:\STAAD Examples\c02_staad_connect* by clicking the button next to the **Location** edit box.

Step 2: Select **Analytical** in the **Type** area and **Metric** in the **Units** area. Choose the **Create** button from the left pane of the user interface screen; the file is loaded. Choose the **Editor** tool from the **Utilities** panel of the **Utilities** tab.

Step 3: In this window, specify the commands, as shown in Figure 2-3.

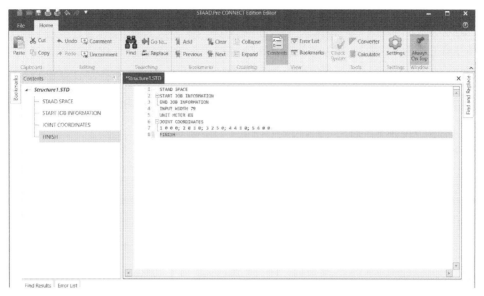

*Figure 2-3 Commands specified in the **Editor** window*

Step 4: Choose the **Save** button from the **File** menu in the **STAAD Editor** window and then close the STAAD Editor; the nodes are generated. Press SHIFT+N to view the node numbers, refer to Figure 2-4.

Figure 2-4 Nodes created using commands

Step 5: Choose the **Close** button from the **File** menu; the current file is closed.

Example 2

In this example, you will create the nodes for a structure using the **REPEAT** and **REPEAT ALL** commands. The structure will be created as a space frame structure.

Steps required to complete this example are given below:

Step 1: Start STAAD.Pro CONNECT Edition and choose the **New** option from the left pane of the user interface screen; the **Model Information** page is displayed. In this page, specify the name *c02_staad_connect_ex2* in the **File Name** edit box and browse to the location *C:\STAAD Examples\c02_staad_connect* by clicking the button next to the **Location** edit box.

Step 2: Select **Analytical** in the **Type** area and **Metric** in the **Units** area. Choose the **Create** button from the left pane of the user interface screen; the file is loaded. Choose the **Editor** tool from the **Utilities** panel of the **Utilities** tab.

Step 3: In the **STAAD Editor** window, specify the commands, as shown in Figure 2-5.

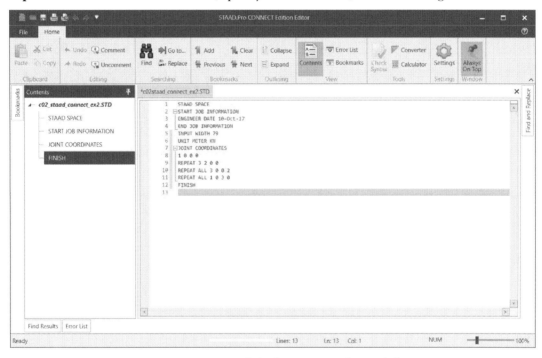

*Figure 2-5 Commands in the **STAAD Editor** window*

Step 4: Choose the **Save** button from the **File** menu in the **STAAD Editor** window and then close the window; the nodes are generated. Press SHIFT+N to view the node numbers, refer to Figure 2-6.

Step 5: Close the file by choosing the **Close** option from the **File** menu.

Note

*Whenever you will save a project in STAAD.Pro, the **REPEAT** and **REPEAT ALL** commands will be converted into comments and the coordinates of each node will be generated in the **STAAD Editor** window.*

Figure 2-6 *Nodes of the space frame structure*

Creating Members

You can create members by specifying their start and end nodes. The command for creating member is **MEMBER INCIDENCES**. The general format for creating members is given below:

```
MEMBER INCIDENCES
m  i  j
```

Here, **m** represents the member number and **i, j** represents the start and end nodes, respectively. After creating nodes, you need to create members by specifying the connectivity between nodes. The **REPEAT** and **REPEAT ALL** commands can also be used for easy generation of the members. The general format to use the **REPEAT** and **REPEAT ALL** commands for creating members is given below:

```
MEMBER INCIDENCES
m  i  j
REPEAT     n iᵢ jᵢ
REPEAT ALL  n iᵢ jᵢ
```

In the above command, **n** represents the number of times, the previously created member will be repeated, i_i represents the member number increment and j_i represents the node number increment.

In this case, the **REPEAT** command will create the previously created member for specified number of times with specified increment in node numbers and member numbers. The **REPEAT ALL** command will create all the previously created members for specified number of times with specified increment in node numbers and member numbers.

Note
*To use the **REPEAT** and **REPEAT ALL** commands for creating members, you need to number the members in a consecutive order.*

Example 3

In this example, you will create the members to form a complete portal frame structure. The file used in this example is *c02_staad_connect_ex1.std*.

Steps required to complete this example are given below:

Step1: Choose the **Open** option from the left pane in the user interface of STAAD.Pro; the **Open** screen is displayed. In this screen, browse to the location: *C:\ STAAD Examples\c02_staad_connect* and select the *c02_staad_connect_ex1.std* file and then choose the **Open** button; the user interface screen is displayed with the nodes. Press SHIFT+N to view the node numbers, refer to Figure 2-7.

Figure 2-7 Node numbers for the nodes

Step 2: Invoke the **STAAD Editor** window and specify the commands for creating the members, as shown in Figure 2-8.

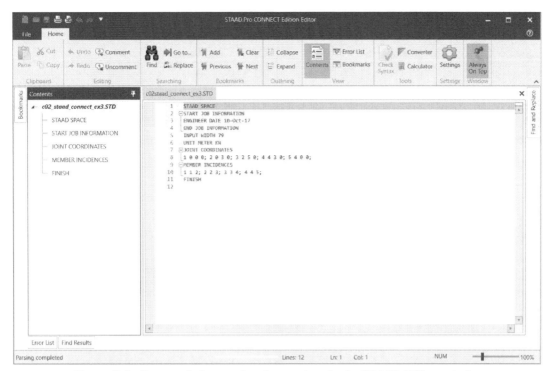

Figure 2-8 *Commands for creating the members in the* ***STAAD Editor*** *window*

Step 3: Choose the **Save** button from the **File** menu in the **STAAD Editor** window and then close the window; the members are generated. Press SHIFT+B to view the member numbers, refer to Figure 2-9.

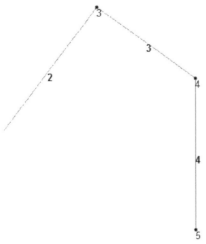

Figure 2-9 *Members created to complete the structure*

Step 4: Choose the **Save As** option from the **File** menu; the **Save As** page is displayed. In this page, specify the name *c02_staad_connect_ex3* in the **File name** edit box and save it at an appropriate location. Choose the **Save** button; the file is saved with the new name.

Step 5: Close the file by choosing the **Close** option from the **File** menu.

Example 4

In this example, you will create members using the **REPEAT** and **REPEAT ALL** commands to form a complete space frame structure. The file used in this example is *c02_staad_connect_ex2.std*.

Steps required to complete this example are given below:

Step 1: Open the *c02_staad_connect_ex2.std* file, as discussed in Example 3. Choose the **Geometry** tab and press SHIFT+N to view the node numbers.

Step 2: Invoke the **STAAD Editor** window and specify the commands given below after the **Joint Coordinates** command:

```
MEMBER INCIDENCES
1 1 17
```

Step 3: After specifying the above command, choose the **Save** button from the **File** menu and close the **STAAD Editor** window; the member 1 will be created, as shown in Figure 2-10. Click on working area. Press SHIFT+B to view the member number.

Figure 2-10 Member 1 created using commands

Step 4: Specify the **REPEAT** command after specifying the command in step 2 in the **STAAD Editor** window, as given below. After specifying the command, choose the **Save** button and close the **STAAD Editor** window; the members are created, as shown in Figure 2-11.

```
REPEAT 3 1 1
```

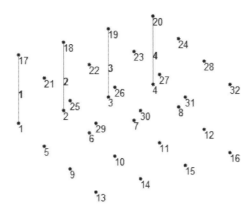

*Figure 2-11 Members created using the **REPEAT** command*

Step 5: Specify the next command given below in the **STAAD Editor** window. Next, choose the **Save** button and close the **STAAD Editor** window; the members are created, as shown in Figure 2-12.

```
REPEAT ALL 3 4 4
```

*Figure 2-12 Members created using the **REPEAT ALL** command*

Step 6: Next, create the horizontal members by specifying the command in the **STAAD Editor** which are given next.

```
17 17 18
REPEAT 2 1 1
REPEAT ALL 3 3 4
```

Step 7: Choose the **Save** button and close the **Editor** window; the members are created, as shown in Figure 2-13.

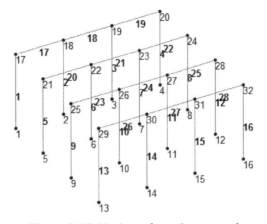

Figure 2-13 Horizontal members created

Step 8: Next, specify the command in the **STAAD Editor** window to create the remaining members as given below and the structure is completed, as shown in Figure 2-14.

```
29 17 21
REPEAT 3 1 1
REPEAT ALL 2 4 4
```

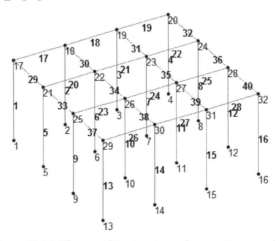

Figure 2-14 The complete structure after creating members

Step 9: Choose the **Save** option from the **File** menu and then close the **STAAD Editor** window.

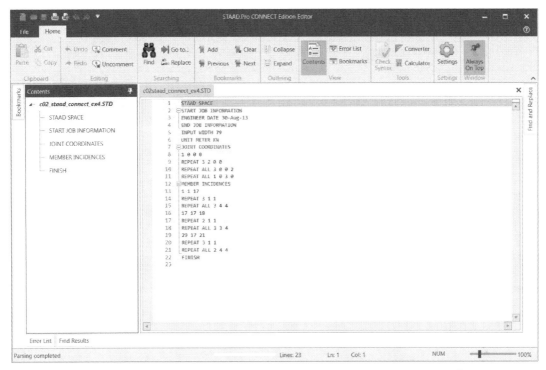

*Figure 2-15 Commands for creating members in the **STAAD Editor** window*

Step 10: Choose the **Save As** option from the **File** menu; the **Save As** page is displayed. In this page, specify the name *c02_staad_connect_ex4* in the **File Name** edit box and save it at an appropriate location.

The commands for creating members can also be written as given below:

```
MEMBER INCIDENCES
1 1 17 4
REPEAT ALL 3 4 4
17 17 18 19
REPEAT ALL 3 3 4
29 17 21 32
REPEAT ALL 2 4 4
```

Step 11: Choose the **Close** option from the **File** menu to close the open file.

In the first line of the above command, the span of the member 1 is from node 1 to node 17 and the new members will be generated with the member number increment and the node number increment as 1. Thus, the spans of the members numbered as 2, 3, and 4 are from nodes 2 to 18, 3 to 19, and 4 to 20, respectively.

Similarly, in the third and fourth line of the command, the members 17 to 19 and 29 to 32 will be formed.

Creating Plate Elements

In **STAAD.Pro**, a plate is a thin shell with multi-noded shape. It can have three or four nodes. A plate is used to model a floor slab, roof, or wall that does not need to be designed as a shear wall. For modeling a shear wall, you need to use the **Add Surface** tool. This tool will be discussed later in this chapter. But, to model a slab, roof, or wall, you need to create several plates which can be done by meshing. The process of meshing will be discussed later in this chapter.

To create a plate, first you need to create nodes. The process of creating nodes is already discussed earlier. Next, identify connectivity between the nodes and create the plates. The command for creating a plate is given next.

```
ELEMENT INCIDENCES
p i₁ i₂ i₃ i₄
```

In the above command, **p** represents plate number, i_1, i_2, i_3, and i_4 represent node numbers. The order of these node numbers can be clockwise or anti-clockwise

If the plate is 3-noded, then i_4 node is not needed. Note that, if you have created members in the project then the member numbers and element numbers must be distinct. You can also create plate elements by using the **REPEAT** and **REPEAT ALL** commands. The general format for using these commands is given below:

```
ELEMENT INCIDENCES
p i₁ i₂ i₃ i₄
REPEAT n eᵢ jᵢ
REPEAT ALL n eᵢ jᵢ
```

In the above command, n represents no. of times the plate will be repeated, e_i represents the element number increment and j_i represents the node number increment.

In this case, the **REPEAT** and **REPEAT ALL** commands will generate the previously created plate n number of times, with the specified plate and node number increment.

Example 5

In this example, you will create plate elements for the plane structure using the **REPEAT** and **REPEAT ALL** commands.

Steps required to complete this example are given below:

Step 1: Create a new file in STAAD.Pro with the name *c02_staad_connect_ex5* and then invoke the **STAAD Editor** window, as discussed earlier. Specify the commands to create nodes and members, as shown in Figure 2-16.

Step 2: Choose the **Save** button and close the **STAAD Editor** window; the nodes and members are created, as shown in Figure 2-17. Press Shift+N and Shift+B to view the no of nodes and beams respectively.

Step 3: Next, invoke the **STAAD Editor** window and specify the command for creating plates, as given below:

```
ELEMENT INCIDENCES
18 1 2 6 5
```

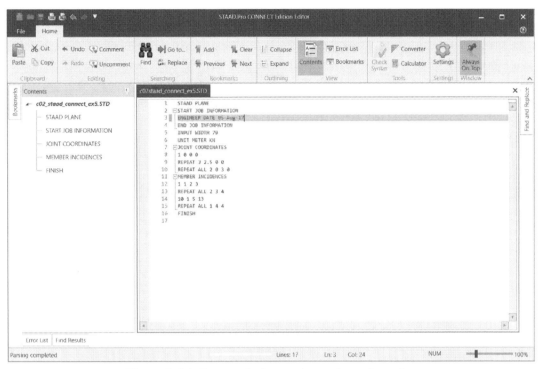

Figure 2-16 Commands for creating nodes and members

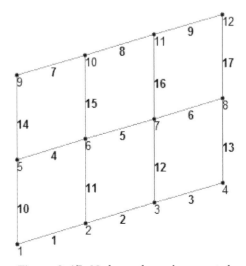

Figure 2-17 Nodes and members created

Step 4: Specify the commands in the **STAAD Editor** window as given below to create the remaining plates, as shown in Figure 2-18. Press SHIFT + P to view the plate numbers.

```
REPEAT 2 1 1
REPEAT ALL 1 3 4
```

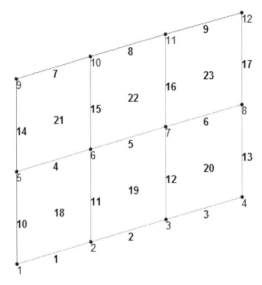

Figure 2-18 *Plates created using commands*

The commands used for creating plates can also be written as given below:

```
ELEMENT INCIDENCES
18 1 2 6 5 TO 20
REPEAT ALL 1 3 4
```

In the first line of the above command, the plate number 18 is formed by connecting the nodes 1, 2, 6, and 5. Plate number 19 and 20 are formed by connecting the node numbers 2, 3, 7, 6 and 3, 4, 8, 7. The node number and plate number increment is 1.

Step 5: Choose the **Save** option from the **File** menu to save the file and then close it by choosing the **Close** option from the **File** menu.

Creating Solid Elements

In STAAD.Pro, solid elements are used to analyze the structures which include 3 dimensional stresses such as concrete gravity dams. A solid element is an eight-noded element and has three translational degrees of freedom per node. To create solid elements, first you need to create the nodes. The command for creating the solid elements is given below:

```
ELEMENT INCIDENCES SOLID
c i₁ i₂ i₃ i₄ i₅ i₆ i₇ i₈
REPEAT nᵢ eᵢ jᵢ
REPEAT ALL nᵢ eᵢ jᵢ
```

Here, **c** is the element number or solid number. i_1 to i_8 represents node numbers, n represents no. of times the solid will be repeated, e_i represents the element number increment and j_i represents the node number increment.

Example 6

In this example, you will create the solid block element in space frame by using the **STAAD Editor** commands.

Steps required to complete this example are given below:

Step 1: Create a new STAAD file with the name *c02_staad_connect_ex6.std* and invoke the **STAAD Editor** window from the **Utilities** panel of the **Utilities** tab. Specify the commands for creating nodes as given below:

```
UNIT METER KN
JOINT COORDINATES
1 0 0 0 3 3 0 0
4 0 0 2 6 3 0 2
REPEAT ALL 1 0 2 0
```

Step 2: Next, specify the commands for creating solid elements as given below:

```
ELEMENT INCIDENCES SOLID
1 4 5 11 10 1 2 8 7 TO 2
```

In the above command, 1 represents the solid number which will be created by connecting the eight nodes in a sequence and TO represents the solid number 2 which will be created automatically by connecting its corresponding nodes in the same manner.

Step 3: Choose the **Save** button and close the **STAAD Editor** window; the solid elements are created, as shown in Figure 2-19. Press SHIFT+C to view the solid element numbers.

Note
*You can view the element in the **Rendered View** window. The **Rendered View** window can be invoked by choosing the **3D Rendering** tool from the **Windows** panel of the **View** tab.*

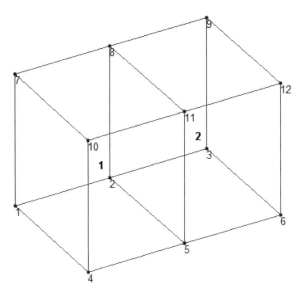

Figure 2-19 *Solid elements created using commands*

Step 4: Choose the **Save** option from the **File** menu to save the file and then close it by choosing the **Close** option from the **File** menu.

STRUCTURAL MODELING USING STAAD GUI

You can model the structural geometry using the STAAD Graphical User Interface (GUI). The STAAD GUI comprises of various graphical tools that are used to generate the structure. Whenever you model the structure using the graphical tools, the **STAAD Editor** window will be automatically updated with the associated commands. Thus, the graphical model generation and the command file methods are interrelated. The changes made in the former will be reflected in the latter and vice-versa. It is essential for the users to learn both the methods so that they can easily make any change and update it in the model. There are various tools and methods available for modeling a structure. Some important methods are discussed next.

Creating Nodes and Members Using Snap Node/Beam Method

In the Snap Node/Beam method, the nodes and members are generated simultaneously. In this method, to create nodes and members, first you need to specify the grid and snap settings. The options related to the grid and snap settings will be available in the **Snap Node/Beam** window. You can invoke the **Snap Node/Beam** window by choosing the **Beam Grid tool** from the **Grids** drop-down in the **Structure** panel of the **Geometry** tab.

In this window, the **Create** button is used to create new grids. The **Edit** button is used to configure the settings of an existing grid. The **Delete** button is used to remove an existing grid from the window. The **Copy** button is used to create a copy of an existing grid. You can change the name of the grid by using the **Rename** button. You can import a grid setting created in AutoCAD by using the **Import** button.

In the **Snap Node/Beam** window, you need to specify the settings of construction lines for creating nodes and members. In STAAD.Pro, there are three grid systems which can be used for specifying the settings of the construction lines. By default, the linear grid system is defined in the window, refer to Figure 2-20. In this window, the active grid system will be selected and highlighted. The three different grid systems available in STAAD.Pro are discussed next.

Linear Grid

In the linear grid system, the construction lines are perpendicular to each other. You can also create a new linear, radial, and irregular grid system. To create a linear grid system, choose the **Create** button in the **Snap Node/Beam** window; the **Linear** dialog box will be displayed, as shown in Figure 2-21.

In the **Linear** dialog box, the **Linear** option will be selected by default in the drop-down list available at the top of the dialog box. You can specify the name of the grid system in the **Name** edit box. Specify the required plane for the grid lines in which the structure will be drawn from the **Plane** area. For example, if a structure is to be drawn in the XY plane, select the **X-Y** radio button. Specify the angle of rotation of a plane about an axis in the corresponding edit box in the **Angle of Plane** area. Next, specify the coordinates of the origin of the grid in the **X**, **Y**, and **Z** edit boxes in the **Grid Origin** area. In the **Construction Lines** area, you

*Figure 2-20 The **Snap Node/Beam** window*

can adjust the settings of the construction lines. To display the coordinates on the negative direction of X and Y axes, specify the required values in the **Left** column of the **X** and **Y** edit boxes by using the spinner. Specify the spacing between the grids in the **Spacing** edit box. To place the axes at an angle, specify the required value in the edit boxes corresponding to **Skew**. Figure 2-22 shows a linear grid system.

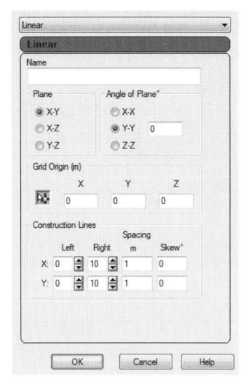

Figure 2-21 The **Linear** *dialog box*   *Figure 2-22* *The linear grid*

Radial Grid

In a radial grid system, the construction lines will appear in a spider-web style. In this grid style, you can create circular structures by creating small linear members. To create a radial grid system, choose the **Create** button from the **Snap/Node Beam** window; the **Linear** dialog box will be displayed, refer to Figure 2-21. Select the **Radial** option from the drop-down list available at the top; the **Radial** dialog box will be displayed and the options related to the radial grid system will be displayed, as shown in Figure 2-23.

In the **Radial** dialog box, you can specify the plane, angle of plane, and grid origin for the radial grid system in the same way as discussed for the linear grid system. In the **Construction Lines** area, specify the start angle of the grid in the **Start Angle** edit box. Specify the total angle of sweep in the **Sweep** edit box. Specify the inner and outer radius in the **Radius 1** and **Radius 2** edit boxes, respectively. Specify the number of bays in the **Bays** edit box. Figure 2-24 shows the radial grid system.

 Note
*Increasing the number of bays in the first **Bays** edit box produces a better circular structure.*

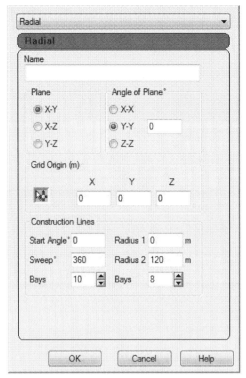

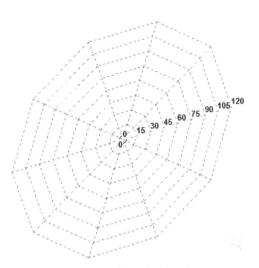

Figure 2-23 The **Radial** dialog box

Figure 2-24 The radial grid system

Irregular Grid

In the irregular grid system, you can create the grid lines with unequal spacing. The process of creating an irregular grid is the same as discussed above for the linear and radial grid system. In this case, you can specify the relative grid distance in the X and Y edit boxes of the **Relative gridline distances** area in the **Irregular** dialog box, refer to Figure 2-25.

After creating the required grid system, choose the **OK** button; the **Snap Node/Beam** window will be displayed. Using this window, you can choose the desired grid system in which you want to work. For example, to create a linear structure, select the check box corresponding to the linear grid system. Next, choose the **Snap Node/Beam** button, if not chosen by default; a plus cursor will appear on the grids displayed in the main window area. Next, to create the nodes, click on the intersection point of the grids; the members will also be created along with the nodes. Figure 2-26 shows an irregular grid system.

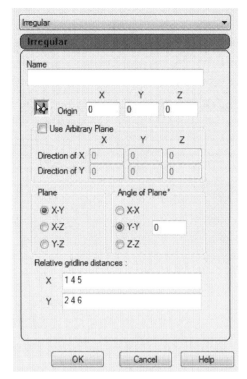

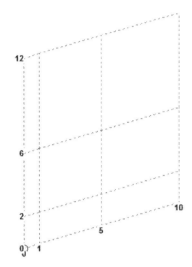

Figure 2-25 The **Irregular** *dialog box*   *Figure 2-26* The **Irregular Grid** *System*

Example 7

In this example, you will create a structure, as shown in Figure 2-27, using the **Snap Node/ Beam** method.

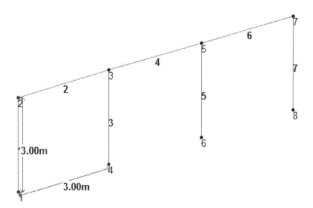

Figure 2-27 Portal frame structure

Steps required to complete this example are given next:

Step 1: Start STAAD.Pro CONNECT Edition and select the **New** option from the left pane; the **Model Information** page is displayed. In this page, specify the name *c02_staad_connect_ex7* in the **File Name** edit box and browse to the location *C:\STAAD Examples\c02_staad_connect* by clicking the button next to the **Location** edit box. Select **Analytical** from the **Type** area and **Metric** from the **Units** area. Next, choose the **Create** button from the left pane of the user interface screen; the file is loaded.

Step 2: Choose the **Beam Grid** tool from the **Grids** drop-down in the **Structure** panel of the **Geometry** tab; the **Snap Node/Beam** window is displayed. In this window, choose the **Edit** button; the **Linear** dialog box is displayed.

Step 3: In this dialog box, ensure that spacing between grids is **1** and other default settings are retained. Choose the **OK** button to close the dialog box.

Step 4: Place the cursor at the origin and click; node1 is created and a member is attached to the cursor, refer to Figure 2-28.

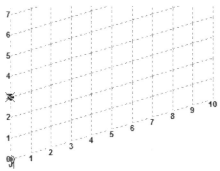

Figure 2-28 Node 1 created at the origin

Step 5: Next, move the cursor in the Y direction and click at (0,3) coordinates; the node 2 is created along with the member 1, refer to Figure 2-29.

Step 6: Now, move the cursor in the X direction and click at the coordinates (3,3); the node 3 is created along with the member 2, refer to Figure 2-29.

Step 7: Move the cursor in the negative Y direction and create node 4 at coordinates (3,0), along with member 3, refer to Figure 2-29.

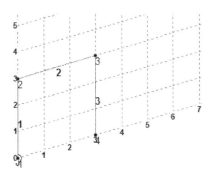

Figure 2-29 Nodes 2, 3, and 4 created along with the members

Step 8: Now, press the ESC key to deactivate the **Snap Node/Beam** mode.

Step 9: Again, choose the **Snap Node/Beam** button from the **Snap Node/Beam** window; the **Snap Node/Beam** mode will be activated and repeat the previous steps to create nodes 5, 6, 7, and 8 with members 4, 5, 6, and 7, refer to Figure 2-27.

Step 10: Next, close the **Snap Node/Beam** window; the structure is created, as shown in Figure 2-27.

Next, press SHIFT + N to show node numbers and SHIFT + B to show beam numbers.

Step 11: Choose the **Save** option from the **File** menu to save the file and then close it by choosing the **Close** option from the **File** menu.

Creating Plate Elements Using the Snap Node/Plate Method

In the Snap Node/Plate method, you can create nodes and plates simultaneously. Before creating the nodes and the plates, first you need to specify the grid and snap settings. To do so, choose the **Quad Plate Grid** tool from the **Grids** drop-down in the **Structure** panel of the **Geometry** tab; the **Snap Node/Plate** window will be displayed, as shown in Figure 2-30. In this window, you can specify the grid and snap settings, which are same as discussed above.

Note
*While working in an existing drawing, the **Snap Node/Plate** window will not be displayed. In that case, choose the **Quad Plate Grid** tool from the **Grids** drop-down in the **Structure** panel of the **Geometry** tab.*

Now, choose the **Snap Node/Plate** button to create the plates. To specify the coordinates for the nodes, click at the appropriate places in the clockwise or anti-clockwise direction in the grid.

Note that the displayed **Snap Node/Plate** window can be used for the quadrilateral plates. To create triangular (3 noded) plates, choose the **Triangular Plate Grid** tool from the **Grids** drop-down in the **Structure** panel of the **Geometry** tab; the **Snap Node/Plate** window for triangular plates is displayed. Now, using this window, you can create triangular (3 noded) plates.

*Figure 2-30 The **Snap Node/Plate** window*

Creating Solid Elements Using the Snap Node/Solid Method

You can create the solid elements by using the Snap Node/Solid method. To do so, choose the **Solid Grid** tool from the **Grids** drop-down in the **Structure** panel of the **Geometry** tab; the **Snap Node/Solid** window is displayed. In this window, specify the required settings for the grid system and choose the **Snap Node/Solid** button; the plus cursor appears in the main window. Next, click at the appropriate places in the grid to create solid elements. Note that, if you click at the random places then the created solid element will be irregular in shape. So, you need to specify the nodes in a proper order. Figure 2-31 shows an irregular solid element which has been created by specifying the nodes randomly.

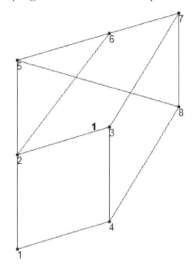

Figure 2-31   An irregular solid element

Example 8

In this example, you will create plate elements using the Snap Node/Plate method. Figure 2-32 shows the structure to be created in this example.

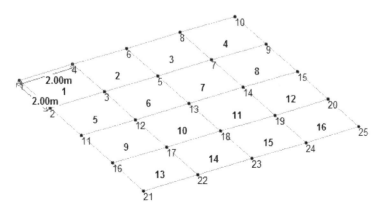

Figure 2-32   The roof slab

Steps required to complete this example are given below:

Step 1: Create a new file in STAAD.Pro with the name *c02_staad_connect_ex8.std* and browse to the location *C:\STAAD Examples\c02_staad_connect* by clicking the button next to the **Location** edit box. Select **Analytical** from the **Type** area and **Metric** from the **Units area**; choose the **Create** button; the user interface screen is displayed. Choose the **Quad Plate Grid** tool from the **Grids** drop-down in the **Structure** panel of the **Geometry** tab. The **Snap Node/Plate** window is displayed in data area.

Step 2: In this window, choose the **Edit** button; the **Linear** dialog box is displayed. In this dialog box, select the **X-Z** radio button in the **Plane** area to activate the XZ plane.

Step 3: Next, enter **8** in the **X** and **Z** edit boxes in the **Right** column edit box/spinner of the **Construction Lines** area and choose the **OK** button to apply the changes.

Step 4: Now, place the cursor at the origin and click; the first node is created.

Step 5: Next, move the cursor in the Z direction and click at 2m; the node 2 is created.

Step 6: Move the cursor in the X direction and click at the coordinates (2,0,2); the node 3 is created.

Step 7: Again, move the cursor in the negative Z direction and click at the coordinates (2,0,0); node 4 and plate 1 is created, as shown in Figure 2-33. Press the ESC key to exit the tool selection. Press SHIFT+N and SHIFT+P to view the nodes and plates respectively.

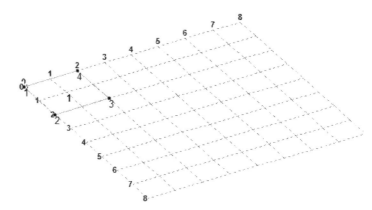

Figure 2-33 Plate 1 created

Step 8: Choose the **Snap Node/Plate** button and then and then click at (4,0,2); node 5 is created.

Step 9: Next, move the cursor in negative Z direction and click at the coordinate (4,0,0); node 6 is created.

Step 10: Move the cursor and click on node 2 and then move the cursor and click on node 3 respectively; plate 2 is created.

Step 11: Repeat the previous steps and create plates 2 to 16, refer to Figure 2-32.

Step 12: Choose the **Save** option from the **File** menu to save the file and then close it by choosing the **Close** option from the **File** menu.

Example 9

In this example, you will create the solid block element using the Snap Node/Solid method. Figure 2-34 shows the concrete block to be created.

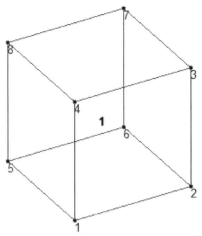

Figure 2-34 *The concrete block*

Steps required to complete this example are given below:

Step 1: Create a new file in STAAD.Pro with the name *c02_staad_connect_ex9.std* and browse to the location *C:\STAAD Examples\c02_staad_connect* by clicking the button next to the **Location** edit box. Select **Analytical** from the **Type** area and **Metric** from the **Units area** in the **Model Information** page and then, choose the **Create** button; the user interface screen is displayed. Choose the **Solid Grid** tool from the **Grids** drop-down in the **Structure** panel of the **Geometry** tab; the **Snap Node/Solid** window is displayed.

Step 2: Move the cursor at the coordinates (0, 0) and click to create node 1.

Step 3: Move the cursor in the X direction and click at the coordinates (4, 0) to create node 2.

Step 4: Similarly, click at the coordinates (4, 4) and (0, 4) to create nodes 3 and 4, respectively.

Step 5: Next, choose the **Edit** button from the **Snap Node/Solid** window; the **Linear** dialog box is displayed.

Step 6: In this dialog box, specify **-4** in the **Z** edit box under the **Grid Origin** area and then choose the **OK** button; the grid is moved to a new location, as shown in Figure 2-35.

Step 7: Now, repeat the steps 2 through 4 to create nodes 5, 6, 7, and 8; the solid element is created, refer to Figure 2-34. Press the ESC button to exit the command. Press SHIFT+N and SHIFT+C to view the node and solid number, respectively.

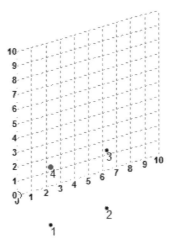

Figure 2-35 *Grids moved to the new location*

Step 8: Choose the **Save** option from the **File** menu to save the file and then close it by choosing the **Close** option from the **File** menu.

STRUCTURAL MODELING USING THE STRUCTURE WIZARD

Structure Wizard contains pre-defined prototype models and templates such as truss models, frame models, surface models, solid models, and so on. Using these templates and prototype models, you can create a structure model by specifying the parameters such as length, width, height, radius, and so on. After creating the model in Structure Wizard, you can transfer or import it into STAAD.Pro to start a project. To access the Structure Wizard, choose the **Structure Wizard** option from the **Structure** panel of the **Geometry** tab; the **default.stp - StWizard** window will be displayed, as shown in Figure 2-36.

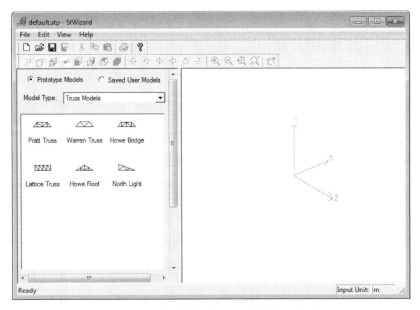

Figure 2-36 *The **default.stp - StWizard** window*

In this window, you can access both type of models, the prototype models and the models saved by the user. To access the prototype models, select the **Prototype Models** radio button. A list of different prototype models will be displayed in the **Model Type** drop-down list such as trusses, plates, solids, surfaces, and so on. To access the saved user models, select the **Saved User Models** radio button. Before generating a structure, you need to specify the units. To do so, choose the **Select Units** option from the **File** menu; the **Select Units** dialog box will be displayed. In this dialog box, select the appropriate unit and choose the **OK** button. The process of generating various types of structures is discussed next.

Truss Models

In the **Structure Wizard** window, you can create truss models. To do so, select the **Truss Models** option from the **Model Type** drop-down list of the **default.stp - StWizard** window; various prototype truss models will be displayed in the left pane of the window. In this pane, select an appropriate truss template. For example, select the **Howe Bridge** truss template and double-click on it; the **Select Parameters** dialog box will be displayed, as shown in Figure 2-37.

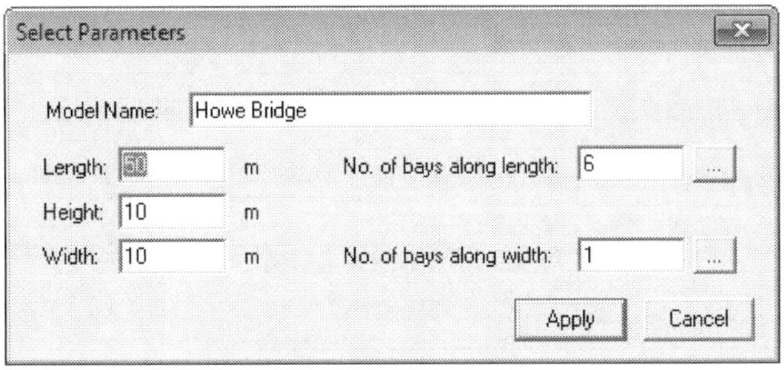

Figure 2-37 *The **Select Parameters** dialog box for **Howe Bridge***

In this dialog box, specify the name for the truss model in the **Model Name** edit box. Next, specify the length, width, height, and no. of bays in their corresponding edit boxes. Choose the **Apply** button; the structure will be generated and displayed in the **Structure Wizard** window. Figure 2-38 shows the Howe bridge model created with the default values.

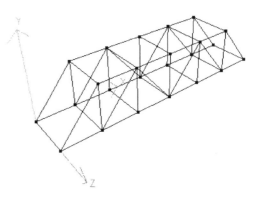

Next, choose the **Merge Model with STAAD.Pro Model** option from the **File** menu; the **StWizard** message box will be displayed. In this message box, choose the **Yes** button; the **Paste Prototype Model** dialog box will be displayed, as shown in Figure 2-39.

Figure 2-38 The Howe Bridge model created

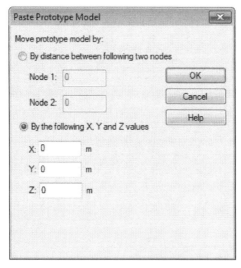

*Figure 2-39 The **Paste Prototype Model** dialog box for a new model*

Note
*The **Paste Prototype Model** dialog box will be displayed only when there is no existing structure in STAAD.Pro. If there is already a structure in STAAD.Pro, then the **Paste Prototype Model** dialog box will be displayed with the **Reference Pt** button added, as shown in Figure 2-40.*

In this dialog box, you can specify the coordinate values to move the prototype model. To move the model by a specified distance, select the **By distance between following two nodes** radio button; the **Node1** and **Node2** edit boxes will be enabled. Specify the desired values and choose the **OK** button to merge the model. To move the model to a specified coordinate, select the **By the following X**, **Y**, **and Z values** radio button; the **X**, **Y**, and **Z** edit boxes will be enabled. Specify the required values and choose the **OK** button to apply the changes. If you merge a new model with an existing one then the **STAAD.Pro CONNECT Edition** message box will be displayed, as shown in Figure 2-41.

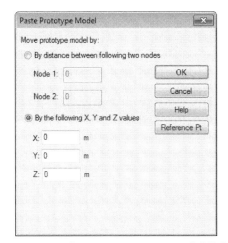

Figure 2-40 The **Paste Prototype Model** *dialog* *Figure 2-41* The **STAAD.Pro CONNECT**
box for an existing model **Edition** *message box*

Choose the **OK** button to close the message box. You can choose the **Reference Pt** button in case of merging one model with other. In that case, you need to specify the node which will act as the reference point. Choose the **OK** button to close the dialog box. On doing so, all the members and nodes will be generated automatically.

Frame Models

You can create the frame models by selecting the **Frame Models** option from the **Model Type** drop-down list in the **Structure Wizard** window. On selecting this option, various prototype frame models will be displayed in the left pane of the window. Using these models, you can create different structures such as continuous beam, bay frame, cylindrical structure, circular beam, floor grid, and so on. In the left pane, double-click on the required prototype model; the **Select Parameters** dialog box for the selected model will be displayed, refer to Figure 2-42. In this dialog box, specify the desired values and choose the **Apply** button; the dialog box closes and the structure is created. Figure 2-43 shows a bay frame structure created with default values.

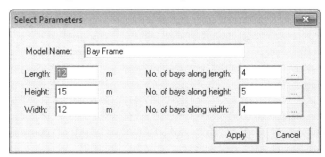

Figure 2-42 The **Select Parameters** *dialog box for the bay frame structure*

Next, you can transfer the model in STAAD.Pro by following the same method as discussed for the truss model.

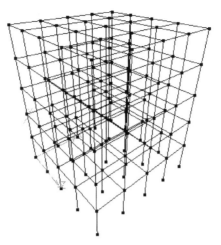

Figure 2-43 The bay frame structure

Surface/Plate Models

Using the **Structure Wizard** window, you can create surface/plates models like quadrilateral plates, cylindrical surfaces, polygonal plates with holes, spherical surface, cooling tower, and so on. To do so, select the **Surface/Plate Models** option from the **Model Type** drop-down list in the **Structure Wizard** window; various prototype models will be displayed in the left pane of the window. In the left pane, double-click on the required model; the dialog box related to the selected model will be displayed. Note that the dialog box displayed for each prototype model will be different. Figure 2-44 shows the dialog box for the **Polygonal Plate With Holes** model.

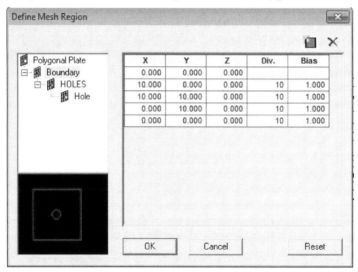

Figure 2-44 The **Define Mesh Region** dialog box for the **Polygonal Plate With Holes** model

In this dialog box, the parameters for the boundary will be displayed by default. You can consider the default values or can change them as per your requirement. Specify the locations for corners of the boundary, number of divisions for each side, and bias for each side division in their respective cells. You can also add a new row in the right pane by choosing the **Add New Row** button available at the top right in the dialog box. To delete a row, first select it and then choose the **Delete Row** button.

Next, click on the **Hole** sub node under the **HOLES** node in the left pane; the corresponding parameters will be displayed in the right pane. Using the options in this pane, you can create circular, polygonal, and elliptical holes. To do so, select the desired option from the **Region Type** drop-down list in the right pane of the dialog box; various parameters will be displayed in the left pane of the dialog box. The options displayed in the dialog box depend upon the options selected from the **Region Type** drop-down list. Specify the values as required and choose the **OK** button; the structure will be created and displayed in the right pane of the window. Figure 2-45 shows the polygonal plate with hole created using the default values. Now, merge the model in STAAD.Pro as discussed before.

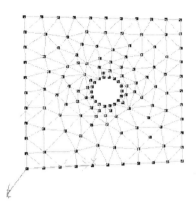

Figure 2-45 *Polygonal plate with hole*

Solid Models

You can also generate solid block models using the Structure Wizard. To do so, select the **Solid Models** option from the **Model Type** drop-down list; the **Solid Block** prototype model will be displayed in the left pane. Double-click on the **Solid Block** option; the **Select Meshing Parameters** dialog box will be displayed, as shown in Figure 2-46.

In this dialog box, specify the length and division along the axes in their corresponding edit boxes. Then, choose the **Apply** button; the solid block is created. Next, transfer it to STAAD.Pro by following the same process as discussed before. Figure 2-47 shows the solid block created using the default values.

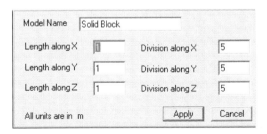

Figure 2-46 *Parial view of the **Select Meshing Parameters** dialog box*

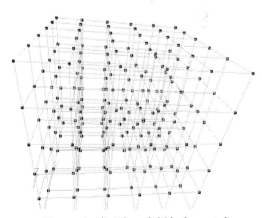

Figure 2-47 *The solid block created*

Composite Models

You can also generate composite models such as bunker or silo using Structure Wizard. To do so, select the **Composite Models** option from the **Model Type** drop-down list; the **Bunker or Silo** prototype model will be displayed in the left pane of the window. Double-click on the **Bunker or Silo** option; the **Select Meshing Parameters** dialog box will be displayed, as shown in Figure 2-48.

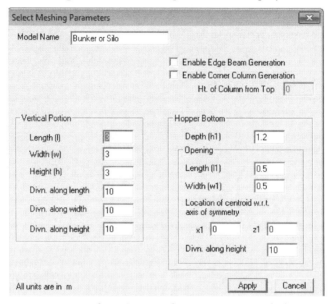

Figure 2-48 *The **Select Meshing Parameters** dialog box*

In this dialog box, you can define the geometry of the vertical portion. To do so, specify the required values in the corresponding edit boxes in the **Vertical Portion** area. Similarly, you can specify the depth of the hopper bottom in the corresponding edit boxes in the **Hopper Bottom** area. After specifying all the parameters, choose the **Apply** button; the structure will be created, as shown in Figure 2-49.

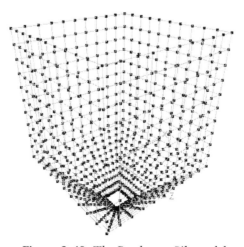

Figure 2-49 *The Bunker or Silo model*

Import CAD Models

Using the **Import CAD Models** option, you can import the AutoCAD models, which are saved in DXF format. To do so, select the **Import CAD Models** option from the **Model Type** drop-down list; the **Scan DXF** and **STAAD Model** options will be displayed in the left pane of the window. To import the AutoCAD models, double-click on the **Scan DXF** option in the right pane; the **Open** dialog box will be displayed. In this dialog box, browse to the required location, select the .dxf file, and then choose the **OK** button; the model will be displayed in the right pane. Using this option, you can import line, 3D-Polyline, and 3D-Face.

To import STAAD models, double-click on the **STAAD Model** option in the right pane; the **Open** dialog box will be displayed. In this dialog box, browse to the required location, select the file, and then choose the **OK** button; the file will be displayed in the window.

VBA Macro Models

Using the **VBA Macro Models** option, you can create models such as stadium roof. To do so, select the **VBA-Macro Models** option from the **Model Type** drop-down list; the **Stadium Roof** and **A Simple Tower** options will be displayed in the left pane of the window. Using these two options, you will be able to model a stadium roof and tower. The procedures to create a stadium roof and a tower are discussed next.

To create a stadium roof, select the **Stadium Roof** option from the left pane of the **default.stp - StWizard** window and drag it to the right pane; the **Stadium Roof** dialog box will be displayed, as shown in Figure 2-50. In this dialog box, you can specify the length, width, depth, number of panels along length and number of panels along width in their corresponding edit boxes. Next, choose the **OK** button; the stadium roof is created and displayed in the window. Figure 2-51 shows the stadium roof model created using default values.

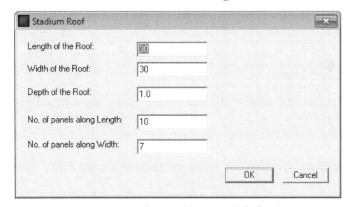

*Figure 2-50 The **Stadium Roof** dialog box*

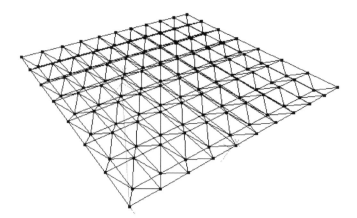

Figure 2-51 Stadium roof created with default values

To create a tower, select the **A Simple Tower** option from the left pane of the **default.stp - StWizard** window and drag it to the right pane; the **Tower Parameters** dialog box will be displayed, as shown in Figure 2-52. In this dialog box, you can specify the base dimension, top dimension, height, and number of bays along height in their corresponding edit boxes. Next, choose the **OK** button; the dialog box is closed and the tower will be created. Figure 2-53 shows a simple tower modeled using default values.

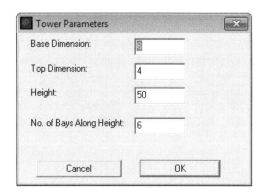

*Figure 2-52 The **Tower Parameters** dialog box*

Figure 2-53 A simple tower created using default values

Example 10

In this example, you will create Howe Bridge in Structure Wizard and transfer it to STAAD.Pro. Figure 2-54 shows the structure of Howe Bridge to be created in this example.

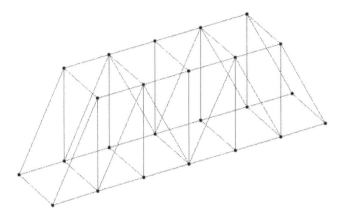

Figure 2-54 *The Howe Bridge model*

Steps required to complete this example are given below:

Step 1: Create a new file in STAAD.Pro with the name *c02_staad_connect_ex10.std* and browse to the location *C:\STAAD Examples\c02_staad_connect* by clicking the button next to the **Location** edit box. Select **Analytical** from the **Type** area and **Metric** from the **Units area** in the **Model Information** page.Next, choose the **Create** button; the user interface screen is displayed. Select **Structure Wizard** from the **Structure** panel of the **Geometry** tab; the **default.stp - StWizard** window is displayed.

Step 2: In this window, ensure that the **Truss Models** option is selected in the **Model Type** drop-down list.

Step 3: Double-click on the **Howe Bridge** option in the left pane; the **Select Parameters** dialog box is displayed.

Step 4: In this dialog box, specify the length, height, width, and number of bays parameters, as shown in Figure 2-55.

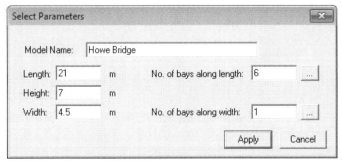

Figure 2-55 *The **Select Parameters** dialog box*

Step 5: Choose the **Apply** button; the model is created and displayed in the window, as shown in Figure 2-56.

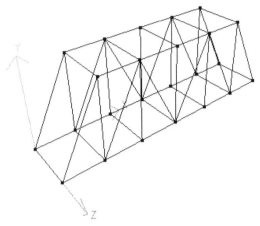

Figure 2-56 *The Howe bridge model*

Step 6: Choose the **Add/Paste Model in STAAD.Pro** option from the **Edit** menu in the **StWizard** window; the **StWizard** message box appears.

Step 7: Choose the **Yes** button; the **Paste Prototype Model** dialog box is displayed. Choose the **OK** button; the model is displayed in the main area of the STAAD.Pro window, refer to Figure 2-54.

Step 8: Choose the **Save** option from the **File** menu to save the file and then close it by choosing the **Close** option from the **File** menu.

Self-Evaluation Test

Answer the following questions and then compare them to those given at the end of this chapter:

1. The command used for creating the nodes is _____.

2. The _____ command repeats the previous line of input specified in the **STAAD Editor** window.

3. In the _____ grid system, the construction lines are perpendicular to each other.

4. The command used for creating solid elements is _____.

5. In STAAD.Pro, plates are used to model slabs and shear walls. (T/F)

6. You can create nodes and members simultaneously using the Snap/Node Beam method. (T/F)

7. In STAAD.Pro, the Structure Wizard contains the prototype models. (T/F)

Review Questions

Answer the following questions:

1. Which of the following commands is used to create members?

    a) **Joint Coordinates**   b) **Element Incidences**
    c) **Member Incidences**   d) None of these

2. Which of the following commands repeats all the previously defined inputs?

    a) **Finish**   b) **Repeat**
    c) **Repeat All**   d) All of the above

3. Which of the following methods is used to create nodes and members simultaneously?

    a) Snap/Node Plate   b) Snap/Node Beam
    c) Snap/Node Solid   d) None of these

4. Which of the following grid styles can be used to create circular structures?

    a) **Radial**   b) **Linear**
    c) **Irregular**   d) All of these

5. In STAAD.Pro, you can create members before creating the nodes. (T/F)

6. Solid elements are the eight-noded elements. (T/F)

7. The **Truss Models** template contains various prototype truss models. (T/F)

Answers to Self-Evaluation Test

1. **Joint Coordinates**, 2. **Repeat**, 3. Linear, 4. **Element Incidences Solid**, 5. T, 6. T, 7. T

Chapter 3

Structural Modeling Using Tools

Learning Objectives

After completing this chapter, you will be able to:

• *Create a portal frame structure*
• *Create a curved beam*
• *Generate finite elements mesh*
• *Create additional structural elements*
• *Generate mirror image of portal frame*
• *Edit the structural elements*

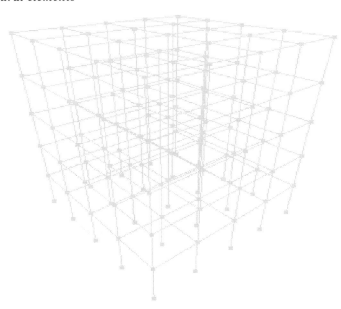

INTRODUCTION

In STAAD.Pro, there are tools which make the modeling process easier and faster. Using these tools, you can add additional nodes, members, plates, and so on. You can also modify an existing structure by using these tools or by modifying the commands. In this chapter, these tools are categorized into two topics which are listed below:

1) Essential tools for structural modeling
2) Other miscellaneous tools

Before using these tools, you need to create nodes and members which has been discussed in the second chapter.

ESSENTIAL TOOLS FOR STRUCTURAL MODELING

STAAD.Pro has some essential tools to create the model geometry. These tools can also be used to modify an existing structure. Using these tools, you can add additional beams, plates, solids, and other structural elements. The procedure for adding beams is discussed next.

Adding Beams Using Tools

In STAAD.Pro, you can add beams to an existing structure using various tools. A beam can also be added to a structure if there are no existing nodes or between the mid-points of the existing members. The process of adding beams and corresponding tools are discussed next.

Add Beam from Point to Point

You can add a beam between two existing nodes in a structure by using the **Add Beam** tool. To do so, invoke the **Add Beam** tool from the **Beam** panel of the **Geometry** tab; the **Add Beam** cursor will be displayed in the user interface. Click on the start node; the beam will be attached to the cursor. Next, click on the end node; the beam will be added.

You can also create a beam when you have an existing member in the drawing but do not have the start and end node for the new beam to be created. To create beams with the help of existing members, choose the **Add Beam** tool from the **Beam** panel of the **Geometry** tab and then click at any point on the member where the start node of the beam will lie; the **STAAD.Pro CONNECT Edition** message box will be displayed prompting you to add a new node. Choose the **Yes** button; the **Insert Nodes into Beam** dialog box will be displayed, as shown in Figure 3-1. In this dialog box, the beam length will be displayed at the top. In the **New Insertion Point** area, specify the distance of the node to be created from the start node in the **Distance** edit box. As you specify the required value in the **Distance** edit box, the proportion of the specified distance with respect to the length of the member will automatically be modified in the **Proportion** edit box. Next, choose the **Add New Point** button; the node location will be added to the **Insertion Points** area.

You can also divide a member into two equal parts by creating a mid-point node. To do so, choose the **Add Beam** tool from the **Beam** panel and select the beam; the **STAAD.Pro CONNECT Edition** message box will be displayed. Choose the **Yes** button from this message box; the **Insert Nodes into Beam** dialog box will be displayed. Choose the **Add Mid Point** button; the mid-point node location will be added to the **Insertion Points** area. Choose the **OK** button; the node will be created.

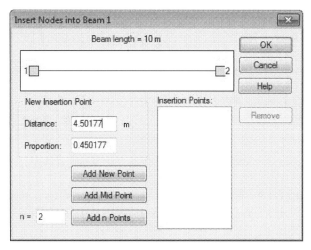

Figure 3-1 The **Insert Nodes into Beam** dialog box

To create a member at the mid point of existing members, choose the **Add Beam** tool from the **Beam** panel and select the beam from where you want to start the new beam; the **STAAD.Pro CONNECT Edition** message box will be displayed. Choose the **Yes** button; the **Insert Nodes into Beam** dialog box will be displayed. Now, choose the **Add Mid Point** button; the mid-point node will be added to the **Insertion Points** area. Choose the **OK** button; the node will be created on the previous beam. Select the new node; the beam will be attached to the node. Next, click at the mid point of the other member where the end node of the beam will be placed; the **STAAD.Pro CONNECT Edition** message box will be displayed prompting you to create a new node. Choose the **Yes** button; the **Insert Nodes into Beam** dialog box will be displayed. Choose the **Add Mid Point** button; the mid-point node will be added to the **Insertion Points** area. Choose the **OK** button; a node will be created. Now. select the new node formed and the beam will be attached at the mid point of another member.

You can also divide beams into any number of equal parts by specifying the required value in the **n** edit box. Next, choose the **Add n Points** button; the location of n number of nodes will be specified in the **Insertion Points** area. To remove a node from the **Insertion Points** area, select the required insertion point and choose the **Remove** button. After specifying the required values, choose the **OK** button, members will be created depending upon the number of nodes inserted.

Note
*You can also insert nodes in an existing member by using the **Insert Node** tool in the **Beam** panel of the **Geometry** tab.*

Add Beam between Mid-Points
Using the **Add Beams between Mid-Points** tool, you can add a beam between the mid-points of two existing members. To do so, choose the **Mid-Point** tool from the **Add Beam** drop-down in the **Beam** panel of the **Geometry** tab. Click on the first member where start node of the beam will lie; the start node of the beam will be created at the mid-point of the member and beam will be attached along with the cursor. Next, click on the member where end node of the beam will lie; the beam will be created between the mid-points of the two existing members.

Add Beam by Perpendicular Intersection

To add a beam between an existing node on a member and the node which is perpendicularly intersecting the other member. To do so, choose the **Perpendicular Intersection** tool from the **Add Beam** drop-down in the **Beam** panel. Next, click on the existing node which will be the start node of the beam. Click on the member which is perpendicular to the beam; a beam will be created perpendicular to the members.

Note
If there is no existing node on the member for the beam to intersect perpendicularly then a node will be created automatically on the member where the beam will intersect.

Add Curved Beam

You can create curved members between two nodes using the **Curved Beam** tool. To do so, select the **Curved Beam** tool from the **Add Beam** drop-down; the curved beam cursor will be displayed in the user interface screen. Next, click on the two nodes; the **Curved Beam Properties** dialog box will be displayed, as shown in Figure 3-2. In this dialog box, specify the radius of curvature and gamma angle in the corresponding edit boxes and choose the **OK** button; the beam will be created according to the specified values.

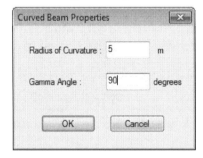

*Figure 3-2 The **Curved Beam Properties** dialog box*

Note
In this chapter, you need to download the c03_staad_connect_ex.zip file for the examples from http://www.cadcim.com. The path of the file is as follows: Textbook > Civil/GIS > STAAD.Pro > Exploring Bentley STAAD.Pro CONNECT Edition.

Example 1

In this example, you will create the portal plane frame structure, shown in Figure 3-3, using the **Add Beam** tool.

Steps required to complete this example are given next.

Step 1: Create a new file in STAAD.Pro with the name *c03_staad_connect_ex1* and create a portal frame structure of **9m*9m** using the **Snap Node/Beam** tool, as shown in Figure 3-4.

Step 2: Invoke the **Add Beam** tool and move the cursor in the user interface screen; the Add Beam cursor is displayed.

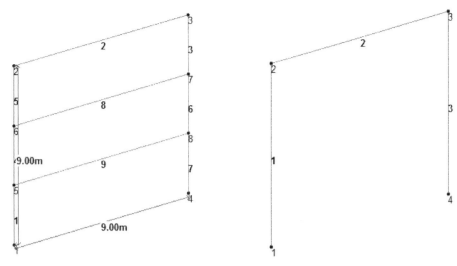

Figure 3-3 The portal frame structure to be created for Example 1

Figure 3-4 The portal frame structure of 9m*9m

Step 3: Next, place the cursor on the member 1 and click; the **STAAD.Pro CONNECT Edition** message box is displayed.

Step 4: Choose the **Yes** button; the **Insert Nodes into Beam 1** dialog box is displayed. In this dialog box, specify the values shown in Figure 3-5.

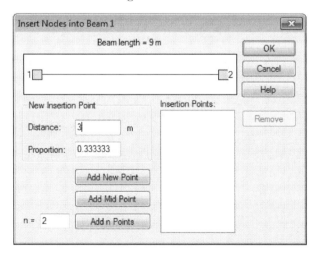

Figure 3-5 The **Insert Nodes into Beam 1** dialog box

Step 5: Choose the **Add n Points** button; the location for the nodes is displayed in the **Insertion Points** area.

Step 6: Next, choose the **OK** button; the nodes 5 and 6 are created.

Step 7: Click on member 3 and repeat the process followed in steps 4 and 5 and create nodes 7 and 8.

Step 8: After creating nodes 7 and 8, click on nodes 6 and 7 to create member 8.

Step 9: Next, click on the nodes 5 and 8 to create member 9.

Step 10: Choose the **Save** button to save the file and close the file by choosing the **Close** option from the **File** menu.

Example 2

In this example, you will create a curved beam using the **Curved Beam** tool.

Steps required to complete this example are given below:

Step 1: Create a new file in STAAD.Pro with the name *c03_staad_connect_ex2* and create a portal frame structure of dimension **6m*6m** using the **Snap Node/Beam** tool, refer to Figure 3-6.

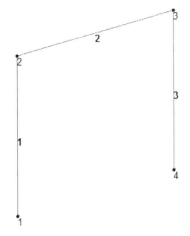

*Figure 3-6 The portal frame structure of 6m*6m*

Step 2: Delete the member 2 and choose the **Curved Beam** tool from the **Add Beam** drop-down in the **Beam panel** of the **Geometry** tab. Move the cursor in the user interface screen; the **Curved Beam** cursor is displayed.

Step 3: Place cursor on node 2 and click; the member is attached with the cursor. Next, click on the node 3; the **Curved Beam Properties** dialog box is displayed.

Step 4: In this dialog box, specify the values, as shown in Figure 3-7 and choose the **OK** button; the curved beam is created, as shown in Figure 3-8.

Step 5: Choose the **Save** button to save the file and close the file by using the **Close** option in the **File** menu.

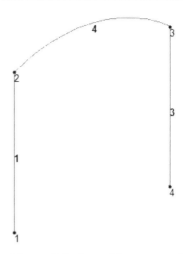

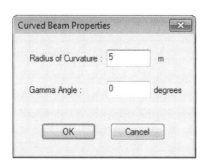

Figure 3-7 The *Curved Beam Properties* dialog box

Figure 3-8 Curved beam created

Creating Collinear Beams

You can create collinear beams (members) along three or more nodes by connecting them along a straight line. To do so, first select the collinear nodes along which members will be created. Next, choose the **Create Collinear Beams** tool from the **Add Beam** drop-down in the **Beam** panel of the **Geometry** tab; the members will be created. Note that if you select more than two nodes then the **STAAD.Pro CONNECT Edition** window will be displayed prompting you about the created members. Choose the **OK** button; the members will be created along the collinear nodes. If there are no collinear nodes then the **STAAD.Pro CONNECT Edition** window will display a message prompting you that no three collinear nodes were found to create new beams.

Creating Beams Along Axes

Members can also be created by connecting collinear points along a selected global axis. To do so, first select the required nodes along which the members will be created. Next, choose the required global axis along which the members will be created from the flyout displayed when you select **Connect Beam** drop-down in the **Beam** panel of the **Geometry** tab; the **STAAD.Pro CONNECT Edition** window will be displayed. Choose the **OK** button; the members will be created along the selected axis. If there are no collinear points along the selected global axis then the **STAAD.Pro CONNECT Edition** window will display a message prompting you that no members are created.

Example 3

In this example, you will create the structure shown in Figure 3-9 using the tools explained above.

Steps required to complete this example are as follows:

Step 1: Open the *c03_staad_connect_ex1.std* file in STAAD.Pro; the model is displayed in the main window, as shown in Figure 3-9.

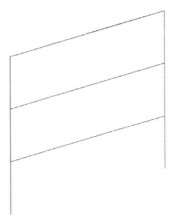

Figure 3-9 *The portal frame structure to be created for Example 3*

Step 2: Select member 2 and right-click; a shortcut menu is displayed. Choose the **Insert Node**; the **Insert Nodes Into Beam 2** dialog box is displayed. Choose the **Add Mid Point** button; the location for the node is displayed in the **Insertion Points** area. Next, choose the **OK** button; the node 9 is created.

Step 3: Select all the beams in the structure and press **Delete**; the **STAAD.Pro CONNECT Edition** message box is displayed, as shown in Figure 3-10. Choose the **OK** button; again the **STAAD.Pro CONNECT Edition** message box is displayed, as shown in Figure 3-11. Choose the **No** button; all beams are deleted, as shown in Figure 3-12. Ensure that the **Geometry** tab is selected.

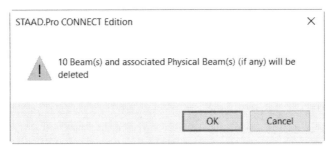

Figure 3-10 *The **STAAD.Pro CONNECT Edition** message box*

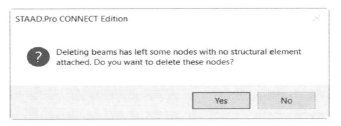

Figure 3-11 *The **STAAD.Pro CONNECT Edition** message box*

Step 4: Invoke the **Nodes Cursor** tool from the **Selection** panel of the **Geometry** tab and select all the nodes; the selected nodes get highlighted.

Step 5: Choose the **Create Collinear Beams** tool from the **Add Beam** drop-down in the **Beam** panel; the **STAAD.Pro CONNECT Edition** window is displayed. Choose the **OK** button; 8 members are created along the collinear points, refer to Figure 3-13.

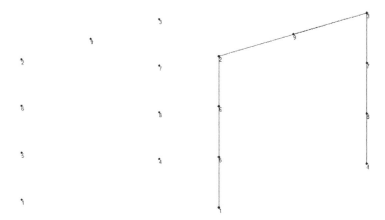

Figure 3-12 *Nodes displayed in the main window after deleting beams*   *Figure 3-13* *Created members*

Step 6: Using the CTRL key, select the nodes numbered 5, 6, 7, and 8 and choose the **X Axis** option from the **Connect Beams** drop-down in the **Beam** panel; the **STAAD.Pro CONNECT Edition** window is displayed. Choose the **OK** button; two members are created along the x axis and the structure gets completed, refer to Figure 3-9.

Step 7: Choose the **Save As** option from the **File** menu; the **Save As** dialog box is displayed. In this dialog box, specify the name *c03_staad_connect_ex3* in the **File name** edit box and save it at an appropriate location. Close the file by using the **Close** option in the **File** menu.

Creating Plates Using Tools

You can add quadrilateral or triangular plates by connecting the existing nodes using certain tools. These tools are discussed next.

To add quadrilateral plates, choose the **Add Plate** tool from the **Plate** panel of the **Geometry** tab. Move the cursor in the user interface screen; the plate icon with the cursor will be displayed. Click on the four nodes in a proper sequence; the plate will be created.

Similarly, to create triangular plates, choose the **Triangular plate** tool from the **Add Plate** drop-down in the **Plate** panel of the **Geometry** tab. The triangular tool is attached with the cursor. Next, click on three nodes to create triangular plates;the plate will be created.

Creating Plates Using Meshing

In STAAD.Pro, meshing can be used to create multiple plates at a time. You can create mesh by using two methods: Generate Surface Mesh and Generate Plate Mesh. These methods are discussed next.

Generate Surface Mesh

The Generate Surface Meshing method is used to generate finite element mesh. In this case, you need to define the outer boundary of the mesh to be generated by connecting the existing nodes. To create surface meshing, choose the **Surface Mesh** tool from the **Generate Mesh** drop-down in the **Plate** panel of the **Geometry** tab; the surface mesh cursor will be attached and displayed. Next, click on the nodes either in the clockwise or anti-clockwise direction; a line will be formed while clicking, which indicates the boundary of the mesh. To close the boundary, click on the first node again. If the boundary is formed by connecting four nodes then the **Choose Meshing Type** dialog box will be displayed, as shown in Figure 3-14.

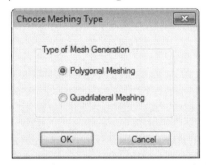

Figure 3-14 The **Choose Meshing Type** *dialog box*

In this dialog box, select the **Polygonal Meshing** radio button to create only triangular elements. Next, choose the **OK** button; the **Define Mesh Region** dialog box will be displayed, as shown in Figure 3-15. Various options in this dialog box are discussed next.

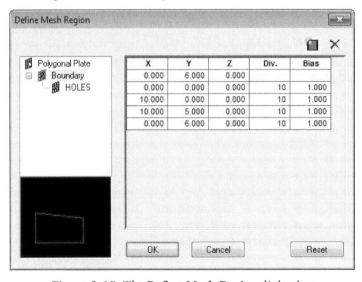

Figure 3-15 The **Define Mesh Region** *dialog box*

In this dialog box, the coordinates of the selected nodes will be displayed in the **X**, **Y**, and **Z** columns in the right pane. In the **Div.** column, you can specify the number of plates to be created along the sides. By default, the value in the **Div.** column is **10**. In the **Bias** column, you can specify a value for the increase or decrease in the size of element along a certain edge. To create a hole in the mesh, click on the **HOLES** node in the left pane in the dialog box and then right-click on it; a shortcut menu is displayed. To create a new hole, choose the **Create New Hole** option; the **Hole 1** will be added under the **HOLES** node and parameters to create a hole will be displayed in the right pane. In this pane, you can define the type of hole to be created by selecting the required option from the **Region Type** drop-down list; the parameters for the hole will be displayed in the right pane. Specify the required values for the option selected and choose the **OK** button; a mesh will be generated.

To create quadrilateral and triangular plates, select the **Quadrilateral Meshing** radio button, refer to Figure 3-14. Next, choose the **OK** button; the **Select Meshing Parameters** dialog box will be displayed, as shown in Figure 3-16.

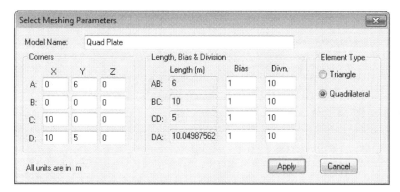

*Figure 3-16 The **Select Meshing Parameters** dialog box*

In this dialog box, you can specify the name of the model, if required, in the **Model Name** text box. The **Quadrilateral** radio button is selected by default in the **Element Type** area. To create triangular plates, select the **Triangle** radio button. The coordinates for the four corners and side lengths are specified in their respective edit boxes. You can specify the required division value in the **Divn** column corresponding to the required sides. After specifying the parameters, choose the **Apply** button; meshing will be created.

If the boundary is formed by connecting 3 nodes or more than 4 nodes then the **Define Mesh Region** dialog box will be displayed. The parameters displayed in this dialog box have been discussed earlier.

Generate Plate Mesh

The Generate Plate Mesh method is used to divide an existing plate into multiple plate elements. To create a plate mesh, invoke the **Plate Cursor** tool from the **Selection** panel of the **Geometry** tab; the plate tool is attached to the cursor. Now, select the required plate and then right click; a shortcut menu is displayed. Choose the **Generate Plate Mesh** option from the menu; the **Choose Meshing Type** dialog box will be displayed along with the **Generating Mesh** dialog box, as shown in Figure 3-17. Select the required radio button from the **Choose Meshing Type** dialog box and choose the **OK** button. The process of creating mesh has been discussed earlier.

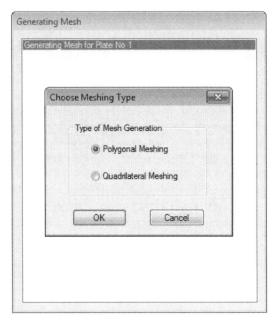

Figure 3-17 The **Choose Meshing Type** and **Generating Mesh** dialog boxes

Creating Infill Plates

In STAAD.Pro, multiple plates can be created at a time by using the **Create Infill Plates** tool. In this case, the plates are generated in a closed boundary surrounded by members. To use this tool, first select the panel which is bounded by members. Next, choose the **Infill Plates** tool from the **Add Plate** drop-down in the **Plate** panel of the **Geometry** tab; the **STAAD.Pro CONNECT Edition** window will be displayed prompting you to create plates. Choose the **OK** button; the plates will be generated. You can view the plates in 3D view.

Example 4

In this example, you will generate finite element mesh to create a roof slab. Figure 3-18 shows the portal frame structure to be created.

Steps required to complete this example are given below:

Step 1: Create a new file in STAAD.Pro with the name *c03_staad_connect_ex4* and create the structure shown in Figure 3-18 using any of the model generation methods discussed in Chapter 2.

Step 2: After creating the structure, choose the **Surface Mesh** tool from the **Generate Mesh** drop-down in the **Plate** panel of the **Geometry** tab. Move the cursor in the Main Window area; the surface meshing cursor appears.

Step 3: Click on the node 2; a line is attached with the cursor. Next, click on node 6, 7, and 3. Click on the node 2 again to close the boundary; the **Choose Meshing Type** dialog box is displayed.

Step 4: Select the **Quadrilateral Meshing** radio button in the dialog box and choose the **OK** button; the **Select Meshing Parameters** dialog box is displayed.

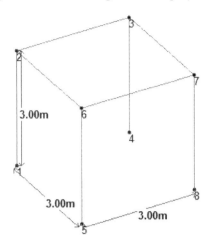

Figure 3-18 *The portal frame structure to be created for Example 4*

Step 5: In this dialog box, specify **6** as the value in the **Divn** column of each side.

Step 6: Next, choose the **Apply** button; the mesh is created. Figure 3-19 shows the created slab. You can also view the generated plates in the rendered view. Press SHIFT+P to view the plate numbers and save the file by choosing the **Save** button.

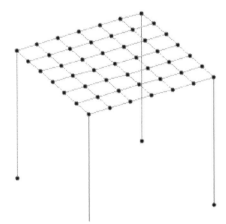

Figure 3-19 *The roof slab created*

Creating Solid Elements Using the Add Solid Tool

The solid elements can be created by connecting the existing nodes in a proper sequence. Solid elements are minimum four to eight noded so you first need to create the nodes. Invoke the **Add Solid** tool from the **Solid** panel of the **Geometry** tab. Now, move the cursor in the user interface screen; the solid cursor will be displayed. Click on the nodes in a proper sequence to connect them; the solid element will be created..

Creating a Structure Using the Translational Repeat Tool

In STAAD.Pro, modeling a tall structure or any complex model is a tedious task. But it can be made simpler by using the **Translational Repeat** tool. Using this tool, you can duplicate and repeat the entire structure or the required portion of the structure multiple number of times in a linear direction by specifying certain parameters, which are discussed next.

To create a copy or multiple copies of an existing structure, first select the structure and then choose the **Translational Repeat** tool from the **Structure** panel of the **Geometry** tab; the **Translation Repeat** dialog box will be displayed, as shown in Figure 3-20.

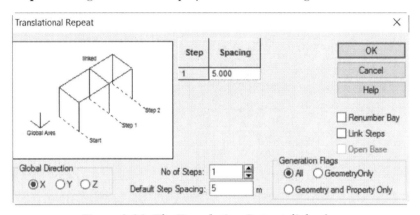

*Figure 3-20 The **Translation Repeat** dialog box*

In this dialog box, to specify the direction in which the structure will be duplicated, select the **X**, **Y**, or **Z** radio button in the **Global Direction** area. Next, specify the number of times the structure will be copied in the **No. of Steps** edit box. As you define the number of steps, the specified number of steps will be added as rows in the **Step Spacing** table. Specify the spacing value between the structure and its copy in the **Default Step Spacing** edit box. You can change the spacing of any step by specifying a value in the **Spacing** column of the **Step Spacing** table. Select the **Renumber Bay** check box to specify the starting number of the members created in each step. On selecting the **Renumber Bay** check box, the **Number From** column will be added to the **Step Spacing** table in which you can specify the desired number. Select the **Link Steps** check box to connect the copied structures by creating members between each copied structure. On selecting the **Link Steps** check box, the **Open Base** check box will be enabled. Select the **Open Base** check box to keep the base of the copied structures open. While generating copies of a structure, either you can duplicate the structure with all its properties and loads acting on it or you can create the copy of the structure only. To duplicate the structure with all its properties and loads, select the **All** radio button in the **Generation Flags** area. To repeat the structure geometry only and not its other properties, select the **GeometryOnly** radio button. To duplicate the structure geometry along with the properties, select the **Geometry and Property Only** radio button. After specifying all the parameters, choose the **OK** button to close the dialog box. All changes will be applied.

Creating a Structure Using the Circular Repeat Tool

Using the **Circular Repeat** tool, you can repeat the entire structure or a small portion of it in a circular direction. To do so, first select the required structure and then choose the

Circular Repeat tool from the **Structure** panel of the **Geometry** tab; the **3D Circular** dialog box will be displayed, as shown in the Figure 3-21. Various parameters displayed in this dialog box are discussed next.

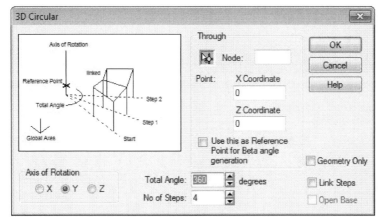

Figure 3-21   *The* **3D Circular** *dialog box*

To specify the axis of rotation for repeating the structure, select the **X**, **Y**, or **Z** radio button in the **Axis of Rotation** area. In the **Through** area, you can specify the coordinates for the axis of rotation. To do so, choose the **Node** button; the node cursor will be displayed in the user interface screen. Next, select the required node in the Main Window area; the coordinates of the selected node will be filled automatically in the **X Coordinate** and **Z Coordinate** edit boxes. Select the **Use this as Reference Point for Beta angle generation** check box to consider the member orientation. Specify the total angle of rotation of the copied structure and original structure in the **Total Angle** edit box. The other parameters in this dialog box are same as for the **Translational Repeat** tool. After specifying all the parameters, choose the **OK** button; the dialog box will be closed and changes will be applied.

Example 5

In this example, you will create a multi-storeyed building using the **Translational Repeat** and **Circular Repeat** tools. Figure 3-22 shows the circular structure to be created in this example.

Figure 3-22   *The circular structure*

Steps required to complete this example are given next.

Step 1: Create a new file in STAAD.Pro with the name *c03_staad_connect_ex5* and create a member using the **Snap Node/Beam** method, refer to Figure 3-23.

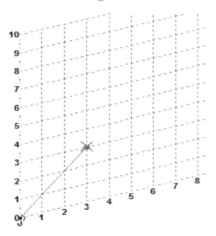

Figure 3-23 Member created using the Snap Node/Beam method

Step 2: Next, close the **Snap Node/Beam** window and select the created member.

Step 3: Choose the **Circular Repeat** tool from the **Structure** panel of the **Geometry** tab; the **3D Circular** dialog box is displayed.

Step 4: In this dialog box, specify the values, as shown in Figure 3-24.

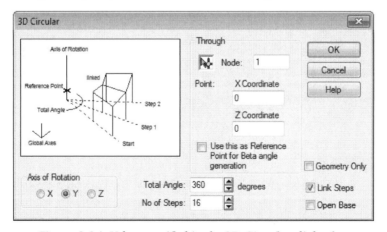

*Figure 3-24 Values specified in the **3D Circular** dialog box*

 Note
*You need to click on the **Node** button to pick the node from the drawing area along which the Circular Repeat will occur.*

Step 5: Next, choose the **OK** button; the dialog box is closed and the member is repeated in the circular direction, refer to Figure 3-25.

Step 6: Select the members 1, 3, 5, 7, 9, 11, 13, 15, 17, 19, 21, 23, 25, 27, 29, and 31, refer to Figure 3-26 and delete them.

Step 7: On deleting all beams, a message box is displayed. Choose the **OK** button; another message box is displayed. Choose the **Yes** button.

Step 8: Invoke the **Beams Cursor** from the **Selection** panel of the **Geometry** tab and select all the members in the user interface screen.

Step 9: Invoke the **Translational Repeat** tool from the **Structure** panel of the **Geometry** tab; the **Translational Repeat** dialog box is displayed, refer to Figure 3-27.

Step 10: Specify the values in the dialog box, as shown in Figure 3-27, and then choose the **OK** button; the dialog box is closed and the structure is created, refer to Figure 3-22. Choose the **Save** button from the **File** tab to save the file. Close the file.

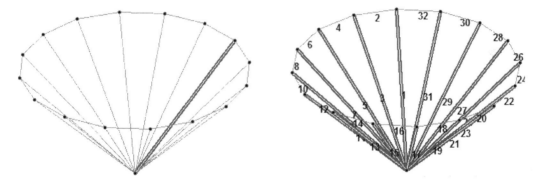

Figure 3-25 *Member repeated in circular direction*   *Figure 3-26* *Selected members*

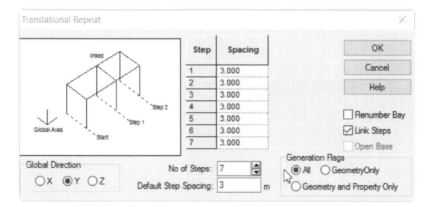

Figure 3-27 *Values specified in the* **3D Repeat** *dialog box*

Shifting Structure Using the Move Tool

In STAAD.Pro, you can shift the entire structure to a new location by specifying the coordinates of that location. Similarly, the location of beams, nodes, and plates can also be changed. This can be achieved by specifying the coordinates in the data area and you can also use the **Move** tool to move the structure which is discussed next.

First select the required nodes, members, plates, or the entire structure. Next, choose the **Move Nodes** tool from the **Move Node** drop-down in the **Node** panel of the **Geometry** tab; a cascading menu will be displayed. In this cascading menu, you can choose the required option. For example, choose the **Move Nodes** option from the cascading menu; the **Move Entities** dialog box will be displayed, as shown in Figure 3-28. You can move the selected entities by specifying the distance between two nodes.

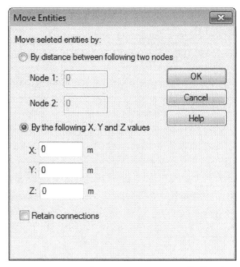

*Figure 3-28 The **Move Entities** dialog box*

To do so, select the **By distance between following two nodes** radio button in the **Move selected entities by** area; the **Node 1** and **Node 2** edit boxes will be enabled. Next, specify the desired node numbers in the **Node 1** and **Node 2** edit boxes. You can also move the selected entities by specifying the required coordinates. To do so, select the **By the following X, Y, and Z values** radio button and specify the desired values in the **X**, **Y**, and **Z** edit boxes. After specifying the values, choose the **OK** button; the dialog box will be closed and the selected entities will move to a new position.

Rotating Structure Using the Generate Rotate Tool

You can rotate an entire structure or a small portion of a structure using the **Rotate** tool. To rotate a structure, first select the required structure and choose the **Generate Rotate** tool from the **Structure** panel of the **Geometry** tab; the **Rotate** dialog box will be displayed, as shown in Figure 3-29. Various parameters displayed in this dialog box are discussed next.

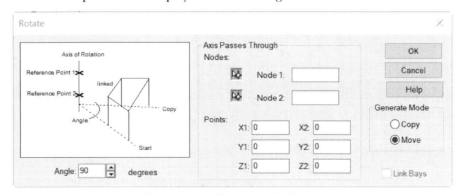

*Figure 3-29 The **Rotate** dialog box*

In this dialog box, specify the angle of rotation in the **Angle** edit box. To define the axis of rotation, choose the **Nodes** button from the **Axis Passes Through** area; the select node cursor will be displayed. Select the required nodes from the user interface screen area; the node numbers will be filled automatically in the **Node 1** and **Node 2** edit boxes in the **Axis Passes Through** area. The coordinates of the selected nodes will also be filled in the **X1**, **Y1**, **Z1**, **X2**, **Y2**, and **Z2** edit boxes in the **Points** area. Select the **Copy** radio button if you want to create a copy of the selected structure. Select the **Move** radio button to save the rotated structure and delete the original one. To connect the original and the copied structures, select the **Link Bays** check box. After specifying all the parameters, choose the **OK** button; the dialog box will be closed and all the changes will be applied to the structure.

Example 6

In this example, you will create a structure using the **Rotate** tool. Figure 3-30 shows the structure to be created in this example.

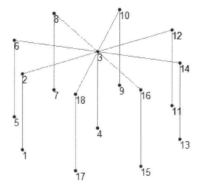

Figure 3-30 *Structure to be created for Example 6*

Steps required to complete this example are given below:

Step 1: Create a new file in STAAD.Pro with the name *c03_staad_connect_ex6* and create a portal plane frame structure of 3m height and 3m width using the **Snap Node/Beam** tool, refer to Figure 3-31. Close the **Snap Node/Beam** window.

Step 2: Select the entire structure and choose the **Generate Rotate** tool from the **Structure** panel of the **Geometry** tab; the **Rotate** dialog box is displayed.

Step 3: Specify the parameters in the dialog box, as shown in Figure 3-32.

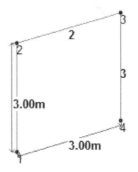

Figure 3-31 *The portal frame structure of 3m*3m*

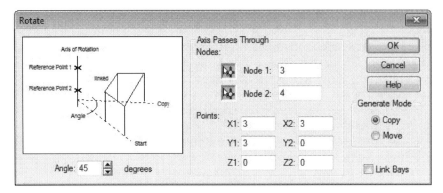

*Figure 3-32 The **Rotate** dialog box*

Step 4: Next, choose the **OK** button; the **STAAD.Pro CONNECT Edition** message box is displayed. Choose the **OK** button from the message box; the structure is rotated and copied, as shown in Figure 3-33.

Step 5: Again, select the entire structure and choose the **Generate Rotate** tool from the **Structure** panel of the **Geometry** tab; the **Rotate** dialog box is displayed. Next, specify the values as mentioned in the dialog box, refer to Figure 3-32. Choose the **OK** button; the **STAAD.Pro CONNECT Edition** message box is displayed. Choose the **OK** button from the message box; the structure is rotated and copied, as shown in Figure 3-34.

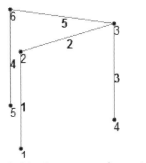

*Figure 3-33 Structure after using the **Rotate** tool*

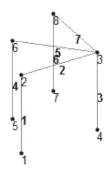

*Figure 3-34 Structure after using the **Rotate** tool*

Step 6: Repeat the procedure followed in steps 5 and create the structure, refer to Figure 3-30. Choose the **Save** button to save the file.

Mirroring a Structure

You can generate a mirror image of the entire structure or a small portion of the structure by using the **Mirror** tool. To do so, first select the required structure and then choose the **Mirror** tool from the **Structure** panel of the **Geometry** tab; the **Mirror** dialog box will be displayed, as shown in Figure 3-35. Various parameters of this dialog box are discussed next.

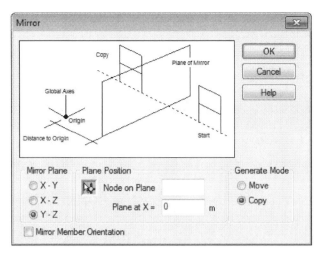

*Figure 3-35 The **Mirror** dialog box*

To define the plane about which the selected geometry will be mirrored, select the radio button corresponding to the required plane from the **Mirror Plane** area. To define the position of the plane, choose the node button and then select the node. Alternatively, specify the node number in the **Node on Plane** edit box. If you want to generate the mirror image of the original structure and do not want the original one, select the **Move** radio button in the **Generate Mode** area. If you want to keep the original structure with the newly created one then select the **Copy** radio button. To mirror the member orientation, select the **Mirror Member Orientation** check box. After specifying all the parameters, choose the **OK** button; the selected geometry will be mirrored about the specified plane.

Example 7

In this example, you will generate the mirror image of a portal frame structure using the **Mirror** tool.

Steps required to complete this example are given below:

Step 1: Create a new file in STAAD.Pro with the name *c03_staad_connect_ex7* and create a portal plane frame structure of **3m*3m** using the **Snap Node/Beam** tool, refer to Figure 3-36.

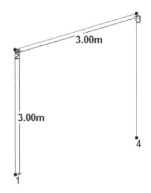

Figure 3-36 *The portal frame structure of 3m*3m for Example 7*

Step 2: Select the entire structure and choose the **Mirror** tool from the **Structure** panel of the **Geometry** tab; the **Mirror** dialog box is displayed.

Step 3: In this dialog box, specify the parameters, as shown in Figure 3-37. Choose the **OK** button; the mirror image is generated, as shown in Figure 3-38.

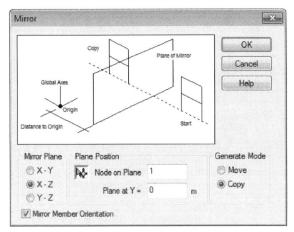

Figure 3-37 *The **Mirror** dialog box*

Figure 3-38 *Mirror image of the structure*

Step 4: Again, select the entire structure and invoke the **Mirror** tool; the **Mirror** dialog box is displayed.

Step 5: In this dialog box, select the **Y - Z** radio button in the **Mirror Plane** area. Specify **3** in the **Node on Plane** edit box in the **Plane Position** area and then select the **Copy** radio button. Next, choose the **OK** button; the mirror image is generated, as shown in Figure 3-39. Choose the **Save** button to save the file. Close the file by using the **Close** option in the **File** menu.

Figure 3-39 *Mirror image of the structure*

OTHER MISCELLANEOUS TOOLS

Earlier in this chapter, you have learned some of the essential tools used for structural modeling. You can also use some of the miscellaneous tools to reduce time and effort while modeling. These tools are discussed next.

Stretching Members Using Stretch Tool

In STAAD.Pro, you can stretch the length of the members in various ways. For example, a member can be stretched through a distance to an existing node, point, or to another existing member. To stretch a member, first select the required member(s) and then choose the **Stretch Beam** tool from the **Beam** panel of the **Geometry** menu; the **Stretch Member(s)** dialog box will be displayed, as shown in Figure 3-40. The options displayed in this dialog box are discussed next.

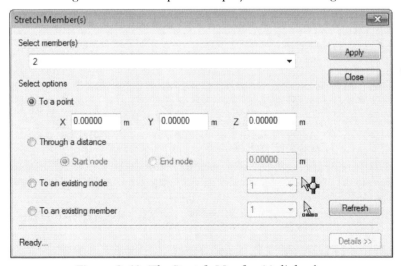

Figure 3-40 *The **Stretch Member(s)** dialog box*

In this dialog box, the selected member(s) will be displayed in the **Select member(s)** drop-down list. If you have selected more than one member then all selected members will be displayed in the drop-down list. You can deselect the selected members by clearing the check box corresponding to the members displayed in the drop-down list. Next, to stretch the selected members, you can use the options available in the **Select options** area which are discussed next.

To stretch the selected members to a point, select the **To a point** radio button and specify the coordinates of the point in the **X**, **Y**, and **Z** edit boxes. Next, choose the **Apply** button; the member(s) will be stretched to a specified point. In that case, STAAD.Pro will determine automatically that which end of the member will be stretched.

To stretch the selected member(s) through a particular distance, select the **Through a distance** radio button. Next, specify the start or end node along which the member will be stretched by selecting the **Start node** or **End node** radio button, respectively. Then, specify the distance through which the member will be stretched in the edit box adjacent to the **End node** radio button. Next, choose the **Apply** button; the selected member(s) will be stretched through the specified distance.

To stretch the selected member(s) to an existing node, select the **To an existing node** radio button. Next, select the required node from the drop-down list available next to the radio button. You can also select the node directly from the Main Window. To do so, choose the **Pick node** button and then move the cursor in the Main Window; the nodes cursor will be displayed. Now, select the node from the Main Window and then choose the **Apply** button; the selected member(s) will be stretched to the selected node.

To stretch the selected member(s) to an existing member, select the **To an existing member** radio button and then select the required member from the drop-down list available next to the radio button. In that case, you can also specify the member by choosing the **Pick member** button.

Note
To stretch a member, there must be an intersection point on the axis of the existing member to be stretched.

Intersecting Members Using the Intersect Tool

Sometimes during the modeling of a structure, the members intersect each other without creating a connection point at the intersection. As a result, there is no transfer of forces between the intersecting members. To avoid this, you can physically connect the intersecting members by creating a common connecting node. To do so, first select the required intersecting members from the user interface screen. Next, choose the **Intersect Selected Beams** tool from the **Intersecting Beams** drop-down in the **Beam** panel of the **Geometry** tab; the **Intersect Members** dialog box will be displayed. In this dialog box, you can specify the tolerance value to find the intersecting members. Next, choose the **OK** button; the **STAAD.Pro CONNECT Edition** message box will be displayed containing information about the newly created members. Choose the **OK** button; new members will be created by splitting the intersecting members at the intersection point.

Merging Members and Nodes

You can merge the collinear members. To merge the members, first select the required members. Next, choose the **Merge Selected Beams** tool from the **Beam** panel of the **Geometry** tab; the

Merge Selected Beams dialog box will be displayed. In this dialog box, the selected members will be displayed and highlighted in blue, refer to Figure 3-41. You can specify the required sectional property, elasticity, Poisson, and density by selecting the required values from the respective drop-down lists. After specifying all properties, choose the **Merge** button; the selected members will be merged into one.

Similarly, you can merge the selected nodes. To do so, select the required nodes and choose the **Merge Nodes** option from the **Node** panel of the **Geometry** tab; the **Select Node** dialog box will be displayed, as shown in Figure 3-42. In this dialog box, select the node to be assigned from the **Node To Keep** drop-down list. Next, choose the **OK** button; the selected nodes will be merged into one.

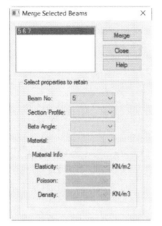

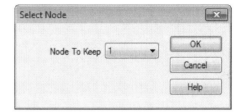

*Figure 3-41 The **Merge Selected Beams** dialog box*   *Figure 3-42 The **Select Node** dialog box*

Renumbering Nodes, Members, and Elements

STAAD.Pro allows you to renumber the nodes, members, plate elements, and solids. To renumber the existing nodes, first select the required nodes from the Main Window. Next, choose the **Renumber Nodes** from the **Node** panel of the **Geometry** tab; the **STAAD.Pro CONNECT Edition** warning message will be displayed prompting you to renumber the nodes. Choose the **Yes** button; the **Renumber** dialog box will be displayed, as shown in Figure 3-43.

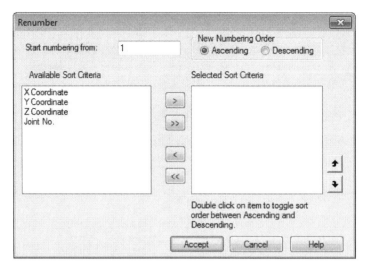

Figure 3-43 *The* **Renumber** *dialog box*

In this dialog box, you can specify the starting number of the node in the **Start numbering from** edit box. Next, specify the numbering order (ascending or descending) by selecting the **Ascending** or **Descending** radio button in the **New Numbering Order** area. You can also specify the sequence of numbering by selecting the required criteria available in the **Available Sort Criteria** area. For renumbering nodes, select the **Joint No.** option and then choose the forward button to shift the selected option in the **Selected Sort Criteria** area. Next, choose the **Accept** button; the nodes will be renumbered. Similarly, you can renumber the members, plates, and solids.

Breaking Beams at Selected Nodes

There are some tools available in STAAD.Pro which allow you to break a beam at selected nodes. If some nodes are created on the line of a member, then you can break the member at those nodes which creates smaller members connecting all the nodes. To break a beam at selected nodes, first select the required node(s) and then select the **Break Beams** tool from the **Beam** panel of the **Geometry** tab; the **STAAD.Pro CONNECT Edition** message box will be displayed informing about the created beams. Choose the **OK** button; the selected beam will be broken at the specified nodes.

CUTTING SECTIONS

Assigning properties and loads to complex structures becomes a tedious task. In that case, you can slice off the structure along the global XY, YZ, and XZ planes to create its sectional view. To create a sectional view, choose the **Cut Section** tool from the **Structure** drop-down in the **Geometry** tools of the **Utilities** tab; the **Section** dialog box will be displayed, as shown in Figure 3-44. This dialog box comprises of three tabs: **Range By Joint**, **Range By Min/Max**, and **Select to View**. These tabs are discussed next.

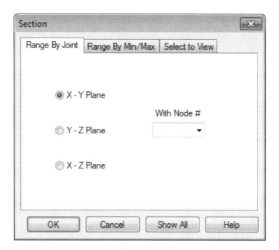

Figure 3-44 The Section dialog box

The **Range By Joint** tab is chosen by default. In this tab, you can specify the section plane by selecting the **X - Y Plane**, **Y - Z Plane**, or **X - Z Plane** radio button. Next, select the node which lie on the section plane from the **With Node #** drop-down list.

In the **Range By Min/Max** tab, you can define a range along the axis perpendicular to the section plane. Thus, the portion of structure lying in this range will be displayed in the Main Window. You can specify a plane by selecting the corresponding radio button. Next, specify the minimum and maximum range for the section in the **Minimum** and **Maximum** edit boxes.

In the **Select to View** tab, you can specify the portion of the structure to be displayed in the Main Window. Select the **Window / Rubber Bound** radio button to select the portion of structure to be displayed by drawing a selection window. To display only the selected objects, select the **View Highlighted Only** radio button. To view only the beams, plates, solids, and nodes, select the **Select to View** radio button. Next, select the corresponding check boxes to display the required object.

Choose the **OK** button from the **Section** dialog box; the required section will be displayed.

After creating the sectional views, you can view the whole structure by choosing the **Whole Structure** option from the **View** tab; the entire structure will be displayed in the main window.

Example 8

In this example, you will use the model given in *c02_staad_connect_ex4.std* and then renumber the nodes and members of the structure.

Steps required to complete this example are given below:

Step 1: Open the *c02_staad_connect_ex4.std* file in STAAD.Pro and press SHIFT+N to view the node numbers.

Step 2: Invoke the **Node Cursor** tool from the **Selection** panel of the **Geometry** tab and select all the nodes. Choose the **Renumber Nodes** tool from the **Node** panel of the **Geometry** tab; the **STAAD.Pro CONNECT Edition** message box is displayed prompting you to proceed with the renumbering process. Choose the **Yes** button; the **Renumber** dialog box is displayed.

Step 3: Next, specify the values in the **Renumber** dialog box, as shown in Figure 3-45.

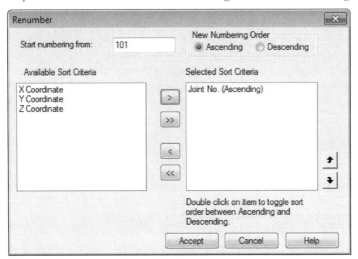

Figure 3-45 *Values specified in the* **Renumber** *dialog box*

Step 4: Choose the **Accept** button; the **STAAD.Pro CONNECT Edition** message box is displayed. Choose the **OK** button; the selected nodes are renumbered.

Step 5: Press SHIFT+N to hide the node numbers and press SHIFT+B to view the member numbers.

Step 6: Select the **Beams Parallel To > X** from the **Beam Cursor** drop-down in the **Selection** panel of the **Geometry** tab; members parallel to **X** axis are selected and highlighted in red in the Main Window.

Step 7: Next, choose the **Renumber Beams** tool from the **Beam** panel of the **Geometry** tab; the **STAAD.Pro CONNECT Edition** message box is displayed. Choose the **Yes** button; the **Renumber** dialog box is displayed.

Step 8: Specify the values in the dialog box, as shown in Figure 3-46.

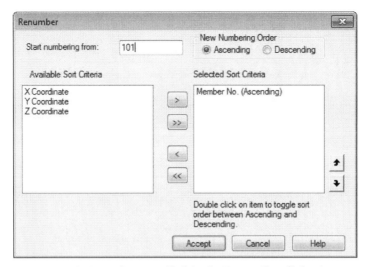

Figure 3-46 *Values specified in the **Renumber** dialog box*

Step 9: Choose the **Accept** button; the **STAAD.Pro CONNECT Edition** message box is displayed prompting you about the renumbered members. Choose the **OK** button; the members are renumbered.

Step 10: Choose **Beams Parallel To > Y** from the **Beam Cursor** drop-down in the **Selection** panel of the **Geometry** tab; beams parallel to Y axis are selected and highlighted in red in the Main Window.

Step 11: Repeat the procedure followed in steps 7 through 9 and renumber the members. In this case, enter **301** in the **Start numbering from** edit box in the **Renumber** dialog box.

Step 12: Choose **Beams Parallel To > Z** from the **Beam Cursor** drop-down in the **Selection** panel of the **Geometry** tab; members parallel to Z axis are selected and highlighted in red in the user interface screen.

Step 13: Repeat the procedure followed in steps 7 through 9 and renumber the members. In this case, enter **401** in the **Start numbering from** edit box of the **Renumber** dialog box.

Step 14: Choose the **Save As** option from the **File** menu; the **Save As** dialog box is displayed. In this dialog box, specify the name *c03_staad_connect_ex8* in the **File name** edit box and save it at an appropriate location. Close the file by using the **Close** option in the **File** menu.

Example 9

In this example, you will use the model given in *c02_staad_connect_ex4.std* and then cut the structure along some of the global planes to view the sections.

Steps required to complete this example are given below:

Step 1: Open the file *c02_staad_connect_ex4.std* in STAAD.Pro and press SHIFT+N to view the node numbers.

Step 2: Choose the **Cut Section** tool from the **Structure tools** drop-down in the **Geometry tools** panel of the **Utilities** tab; the **Section** dialog box is displayed. Now, you will specify the values to view the inside section of the structure.

Step 3: In the **Range By Joint** tab of the dialog box, select the **Y-Z Plane** radio button and node number **2** from the **With Node#** drop-down list.

Step 4: Next, choose the **OK** button; the dialog box closes and the section is displayed, as shown in Figure 3-47.

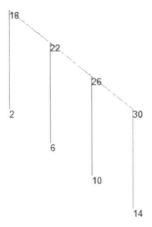

Figure 3-47 Section after cutting

Step 5: Now, choose the **Display Whole Structure** tool from the **Tools** panel of the **View** tab; the whole structure is displayed.

Step 6: Repeat the previous steps to cut off the section along the other global planes.

Step 7: Choose the **Save As** option from the **File** menu; the **Save As** dialog box is displayed. In this dialog box, specify the name *c03_staad_connect_ex9* in the **File Name** edit box and save it at an appropriate location. Close the file by using the **Close** option in the **File** menu.

Self-Evaluation Test

Answer the following questions and compare them to those given at the end of this chapter:

1. The _____ tool is used to add beam between two nodes.

2. The _____ tool is used to add curved beams between two nodes.

3. The _____ tool is used to create surface meshing.

4. The collinear members can be created by using the **Create Colinear Beams** option. (T/F)

5. In STAAD.Pro, only quadrilateral plates can be created. (T/F)

6. In STAAD.Pro, you cannot divide a single plate into multiple plates. (T/F)

7. The **Translational Repeat** tool is used to repeat the structural elements in a circular direction. (T/F)

Review Questions

Answer the following questions:

1. Which of the following tools is used to repeat the structure multiple times in a linear direction?

    (a) **Rotate**   (b) **Circular Repeat**
    (c) **Translational Repeat**   (d) **Copy**

2. Which of the following tools is used to connect intersecting members?

    (a) **Mirror**   (b) **Rotate**
    (c) **Intersect**   (d) None of these

3. Which of the following tools is used to split a member into two parts?

    (a) **Insert Node**   (b) **Split Beam**
    (c) **Move**   (d) **Add Beam**

4. In STAAD.Pro, only nodes can be renumbered. (T/F)

5. Using the **Move** tool, you can shift the structure to a new position. (T/F)

6. Openings can be inserted only in surfaces. (T/F)

7. You can create multiple plates at a time by using the **Create Infill Plates** tool. (T/F)

Answers to Self-Evaluation Test

1. Add Beam from Point to Point, 2. Add Curved Beam, 3. Generate Surface Meshing, 4. T, 5. F, 6. F, 7. F

Chapter 4

Defining Material Constants and Section Properties

Learning Objectives

After completing this chapter, you will be able to:
- *Define material constants and cross-section properties*
- *Define and assign section properties*

INTRODUCTION

After modeling a structure, you need to define its material constants and cross-section properties. In STAAD.Pro, there are two types of materials: Isotropic and Orthotropic 2D. The Isotropic materials have same values of a property in all directions whereas in Orthotropic 2D materials, properties vary in primary direction. The process of defining and assigning constants and cross-section properties is discussed next.

MATERIAL CONSTANTS

The material constants include elastic constants such as Poisson's ratio, Young's modulus, and Shear modulus as well as density, thermal coefficient, and critical damping ratio. You can define the material constants in the **Material** page of the **Specification** tab. In this page, the **Material - - Whole Structure** window will be displayed in the user interface screen, as shown in Figure 4-1.

Figure 4-1 *The **Material - - Whole Structure** window*

In this window, the **Isotropic** tab is chosen by default. In this tab, four pre-defined materials are displayed. To view the values of different properties of a material, double-click on the corresponding material. For example, if you double-click on **CONCRETE**, the **Isotropic Material** dialog box will be displayed, as shown in Figure 4-2. In this dialog box, the name of the material is displayed in the **Title** edit box of the **Identification** area. The material properties and its values are displayed in the **Material Properties** area. You cannot edit the values displayed in this dialog box except the values of compressive strength.

You can use the options displayed in the **Isotropic** tab of the **Material - - Whole Structure** window to create new materials, edit existing materials, change the values of material properties, and to assign the material properties to the structure. This is discussed next.

Figure 4-2 *The **Isotropic Material** dialog box*

Creating New Materials

You can create a new material and define its properties. To do so, choose the **Create** button in the **Material - - Whole Structure** window; the **Isotropic Material** dialog box will be displayed, as shown in Figure 4-3. In this dialog box, you can specify a name for the material in the **Title** edit box. In the **Material Properties** area, specify values of the properties in their respective edit boxes. Next, choose the **OK** button; the **Isotropic Material** dialog box will be closed and the name of the newly created material will be displayed in the **Title** area of the **Material - - Whole Structure** window. You can also create a material tag from the available pre-defined materials. To do so, click on the down-arrow in the **Title** edit box of the **Isotropic Material** dialog box; a list of material names will be displayed. Select the required material from the displayed list; the values of the properties of the selected material will be automatically filled in the edit boxes corresponding to them in the **Material Properties** area, refer to Figure 4-3. Note that these values will not be in the editable mode. To change the values of the properties of existing material, specify a new name in the **Title** edit box; the material properties will be enabled for editing. Now, you can enter the desired property values in the edit boxes and close the dialog box by choosing the **OK** button.

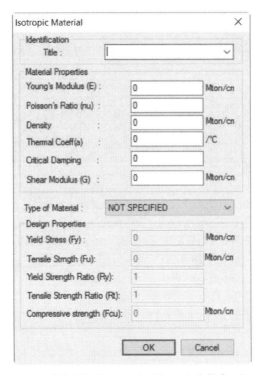

Figure 4-3 The **Isotropic Material** *dialog box*

Editing Material Properties

You can edit the newly created material properties. To do so, first select the material from the **Title** area and then choose the **Edit** button in the **Material - - Whole Structure** window; the **Isotropic Material** dialog box will be displayed. In this dialog box, specify the property values as required and choose the **OK** button to apply the changes. You can use the **Delete** button to delete any material from the materials available in the **Title** area.

Assigning Materials to the Structure

You can assign the required material to the structure. To do so, you can use any of the radio buttons available in the **Assignment Method** area of the **Material - - Whole Structure** window. These radio buttons are discussed next.

Assign To Selected Beams

This radio button is used to assign the desired material to the members selected in the structure. For assigning material to the structural members, you first need to select the required material from the **Title** area. Next, select the structural members or elements from the main window. The **Assign To Selected Beams** radio button is selected by default in the **Assignment Method** area. Now, choose the **Assign** button; the **STAAD.Pro CONNECT Edition** message box will be displayed. Choose the **Yes** button; the **STAAD.Pro CONNECT Edition** message box will close and the material will be assigned to the selected structural members.

Assign To View

This radio button is used to assign a material to an entire structure. For assigning material to the whole structure, first select the desired material and then select the **Assign To View** radio button. Now, choose the **Assign** button; the **STAAD.Pro CONNECT Edition** message box will be displayed. Choose the **Yes** button; the **STAAD.Pro CONNECT Edition** message box will close and the material will be assigned to the entire structure.

Use Cursor To Assign

This radio button is used to assign material to a structural element by using the cursor. For assigning material to a structural element one by one, first select the material and then select the **Use Cursor To Assign** radio button. Next, choose the **Assign** button; the **Assigning** label will appear on it. Now, click on the structural elements to which you want to assign the material. Choose the **Assigning** button to terminate the assigning process.

Assign To Edit List

This radio button is used to assign material by specifying a list of members to which material will be assigned. For assigning material to the structural members, first select the material and then select the **Assign To Edit List** radio button; the edit box below the radio button will be enabled. In this edit box, specify the member numbers and then choose the **Assign** button; the material will be assigned to the specified members.

After assigning material constants, you can view the commands for the defined material constants. To do so, choose the **Editor** button from the **Utilities** panel of the **Utility** tab; the **Warning** message box will be displayed. In this message box, choose the **Save** button; the **Editor** window will be displayed. The commands for the material constant will be displayed below the **Member Incidences** command. For example, if you have assigned concrete material to the structure then the command for the concrete material will be as follows:

```
DEFINE MATERIAL START
ISOTROPIC CONCRETE
E 2.17185e+007
POISSON 0.17
DENSITY 23.5616
ALPHA 1e-005
DAMP 0.05
END DEFINE MATERIAL
CONSTANTS
MATERIAL CONCRETE ALL
```

Defining Orthotropic Material

You can define the material property of the Orthotropic 2D materials as well. To do so, choose the **Orthotropic 2D** tab in the **Material -- Whole Structure** window. In this tab, choose the **Create** button; the **2-D OrthoTropic Material Property** dialog box will be displayed, as shown in Figure 4-4. In this dialog box, specify the name of the material in the **Title** edit box. Specify the values for **Young's Modulus** and **Thermal Coefficient** in local x and y directions in the **Property in Element X direction** and **Property in Element Y direction** areas. Specify the values for **Density**, **Critical Damping** and **Poisson's Ratio** in their corresponding edit boxes in the **General** area. You can specify the values for shear modulii in the **Gxy**, **Gyz** and **Gzx** edit boxes from the **Shear Modulii** area. After specifying the values, choose the **Add** button to add material properties to the **Material - - Whole Structure** window.

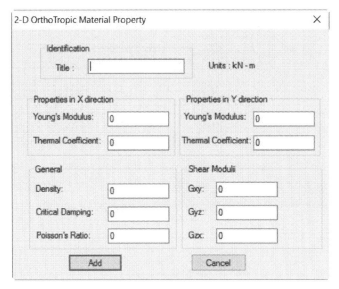

*Figure 4-4 The **2D OrthoTropic Material Property** dialog box*

 Note
For doing the examples of this chapter, you need to download the c04_staad_connect.zip file from http://www.cadcim.com. The path of the file is as follows: Textbook > Civil/GIS > STAAD.Pro > Exploring Bentley STAAD.Pro CONNECT Edition.

Example 1

In this example, you will open the file *c02_staad_connect_ex4* file. Next, you will create a material and assign it to a structure.

The steps required to complete this example are given below:

Step 1: Start STAAD.Pro and open the file *c02_staad_connect_ex4*; the model is displayed in the main window, refer to Figure 4-5.

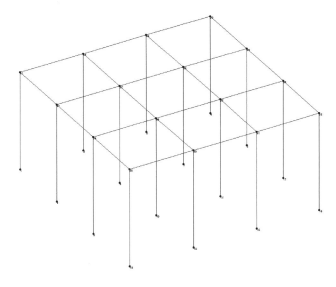

Figure 4-5 *Model displayed in the Main Window*

Step 2: Select all the members and choose the **Translational Repeat** tool from the **Structure** panel of the **Geometry** tab; the **Translational Repeat** dialog box is displayed. Specify the values in this dialog box, as shown in Figure 4-6. Next, choose the OK button; the members are repeated, as shown in Figure 4-7.

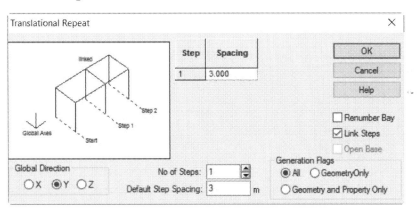

Figure 4-6 The **Translation Repeat** *dialog box*

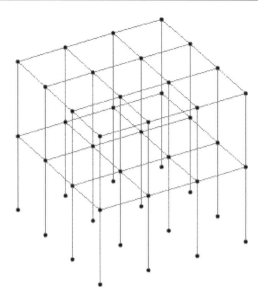

Figure 4-7 *Model displayed in the Main Window*

Step 3: Choose the **Specification** tab and then open the **Material** page from it; the **Material - - Whole Structure** window is displayed.

Step 4: In this window, choose the **Create** button; the **Isotropic Material** dialog box is displayed.

Step 5: In this dialog box, select the **CONCRETE** option from the **Title** drop-down list in the **Identification** area.

Step 6: Now specify a new name **CONC** in the **Title** edit box and specify the values for the properties, as shown in Figure 4-8.

Figure 4-8 The *Isotropic Material* dialog box

Step 7: Choose the **OK** button; the **CONC** material is created and displayed in the **Title** area of the **Isotropic** tab.

Step 8: Select the **CONC** material and then select the **Assign To View** radio button in the **Assignment Method** area. Next, choose the **Assign** button; the **STAAD.Pro CONNECT Edition** message box is displayed. Choose the **Yes** button; the material gets assigned to the entire structure.

Step 9: Choose the **Save As** option from the **File** menu; the **Save As** dialog box is displayed. In this dialog box, specify the name *c04_staad_connect_ex1.std* in the **File name** edit box and save it at an appropriate location.

SECTION PROPERTIES

After assigning the materials to the structure, you need to define the section properties for the structure and then assign those properties to the structure. In STAAD.Pro, you can define the section properties for prismatic sections, steel sections, tapered sections, steel joist, and joist girders from the available database in the software. This is discussed next.

Prismatic Section

You can define the prismatic section properties in the **Properties** page of the **Specification** tab. In this page the **Properties - Whole Structure** window will be displayed in the Data Area, refer to Figure 4-9. In this window, the **Section** tab is chosen by default. In this tab, choose the **Define** button; the **Property** dialog box is displayed with the **Circle** page selected by default, refer to

Figure 4-10. For the circular section, specify the diameter in the **YD** edit box. You can define the rectangular section as well. To do so, choose the **Rectangle** option in the left pane of the dialog box; the **Rectangle** page will be displayed. In this page, specify the length and width of the rectangle in the **YD** and **ZD** edit boxes, respectively.

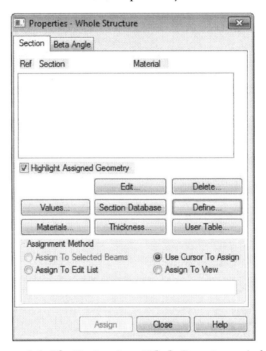

Figure 4-9   The *Properties - Whole Structure* window

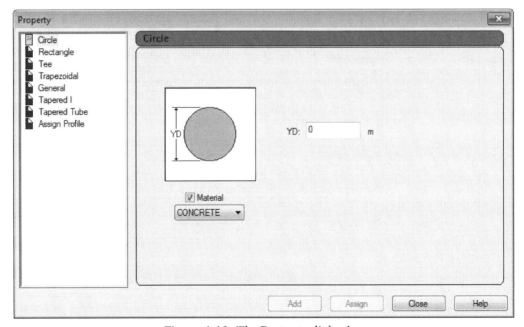

Figure 4-10   The *Property* dialog box

To define the Tee section, choose the **Tee** option; the **Tee** page will be displayed. Next, specify the dimension of the web and flange in the **YD**, **ZD**, **YB**, and **ZB** edit boxes.

To define the trapezoidal section, choose the **Trapezoidal** option; the **Trapezoidal** page will be displayed. Next, specify values in the **YD**, **ZD**, and **ZB** edit boxes to define the geometry of trapezoid.

To define an arbitrary section, choose the **General** option; the **General** page will be displayed. In this page, specify the value for the cross-sectional area, the shear area, and the moment of inertia in their corresponding edit boxes. After specifying the dimensions for the required section, you need to include the material constants. To do so, select the **Material** check box in the right pane of the **Property** dialog box if it is not selected by default. Then, select the required material from the drop-down list available below the check box. Next, choose the **Add** button; the defined section property will be added to the **Section** tab with the tag number and the property description. Now, you can assign the defined section to the structure by using any of the four assignment methods discussed earlier.

You can also view the commands for the section properties assigned to the structure. To do so, choose the **Editor** button from the **Utilities** panel of the **Utilities** tab; the **Warning** message box will be displayed. Choose the **Save** button; the **STAAD Editor** window will be displayed with all the commands. The commands for creating the prismatic rectangular section is given next.

```
MEMBER PROPERTY
1 TO 7 PRIS YD 0.25 ZD 0.35
```

Note

*The commands for the section properties will be displayed after the **Material Constants** command in the **Editor** window.*

After assigning cross-sections to the structure, you can view the rendered view in the **Rendered View** window. To do so, choose the **3D Rendering** button from the **Windows** panel of the **View** tab; the **Rendered View** window will be displayed. Figure 4-11 shows the rendered view of the structure.

Figure 4-11 Rendered view of the structure

Example 2

In this example, you will open the *c04_staad_connect_ex1.std* file in STAAD.Pro. Next, you will define a prismatic section and then assign it to the structure.

Steps required to complete this example are given below:

Step 1: Start STAAD.Pro and open the file *c04_staad_connect_ex1.std*; the model is displayed in the user interface screen.

Step 2: Choose the **Properties** tab and the **Properties - Whole Structure** window are displayed in the user interface screen.

Step 3: In the **Properties - Whole Structure** window, choose the **Define** button; the **Property** dialog box is displayed.

Step 4: In this dialog box, choose the **Rectangle** option from the left pane; the **Rectangle** page is displayed. In this page, specify the values **0.250** and **0.350** in the **YD** and **ZD** edit boxes, respectively. Next, select the material **CONC** from the **Material** drop-down list.

Step 5: Choose the **Add** button to add the rectangular section and close the dialog box. Next, select the added section from the **Section** tab of the **Properties - Whole Structure** window.

Step 6: Select the **Assign To View** radio button from the **Assignment Method** area and then choose the **Assign** button; the **STAAD.Pro CONNECT Edition** message box is displayed. Choose the **Yes** button; the section is assigned to the entire structure.

Step 7: Right-click in the main window and choose the **Structure Diagrams** option from the shortcut menu; the **Diagrams** dialog box is displayed.

Step 8: In this dialog box, select the **Full Sections** radio button from the **3D Sections** area and choose **Apply** and then the **OK** button to close the dialog box. Figure 4-12 shows the rendered view of the structure.

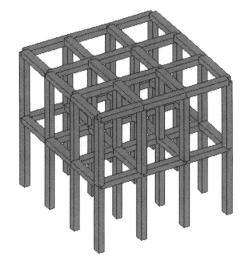

Figure 4-12 *The rendered view of the structure*

Step 9: Choose the **Save As** option from the **File** menu; the **Save As** dialog box is displayed. In this dialog box, specify the name *c04_staad_connect_ex2.std* in the **File name** edit box and save it at an appropriate location. Close the file using the **Close** option in the **File** menu.

Tapered Sections

You can define the tapered I and tube section properties and assign them to a structure. The tapered I section has varying depth along the length of the member. To define the tapered I section, choose the **Define** button from the **Properties - Whole Structure** window and invoke the **Property** dialog box. In this dialog box, choose the **Tapered I** option; the **Tapered I** page will be displayed, as shown in Figure 4-13.

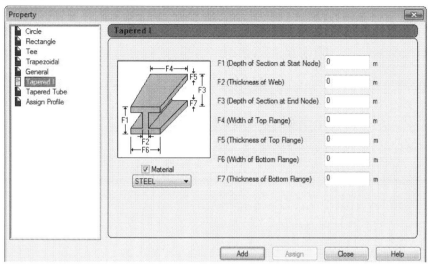

*Figure 4-13 The **Tapered I** page in the **Property** dialog box*

In this page, specify the dimension of the web and flange in the **F1 (Depth of Section at Start Node**, **F2 (Thickness of Web)**, **F3 (Depth of Section at End Node)**, **F4 (Width of Top Flange)**, **F5 (Thickness of Top Flange)**, **F6 (Width of Bottom Flange)**, and **F7 (Thickness of Bottom Flange)** edit boxes. You can include the material constant by selecting the **Material** check box. Next, choose the **Add** button to add the material and then assign it to the structure.

Similarly, to add the tapered tube section properties, choose the **Tapered Tube** option from the left pane of the dialog box; the **Tapered Tube** page will be displayed, as shown in Figure 4-14.

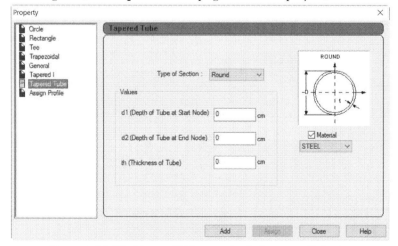

*Figure 4-14 The **Tapered Tube** page in the **Property** dialog box*

In this page, select the type of tube from the **Type of Section** drop-down list. Next, specify the depth of the tube at the start and end nodes in the **d1** and **d2** edit boxes, respectively. Specify the thickness of tube in the **th** edit box. Now, choose the **Add** button to add the structure to the **Section** tab and then close the dialog box by choosing the **Close** button. After adding the section, you can assign it to the structure.

To assign a profile to the structure, select **Assign Profile** from the left pane of the **Property** dialog box; the **Assign Profile** page will be displayed, as shown in Figure 4-15. You can assign the desired profile by selecting the radio button corresponding to that option from the **Select Profile Specification** area and then choosing the **Add** button. On doing so, the selected profile will be added to the **Properties - Whole Structure** window in the **Section** tab. Once a section is added, it can be assigned to the structure.

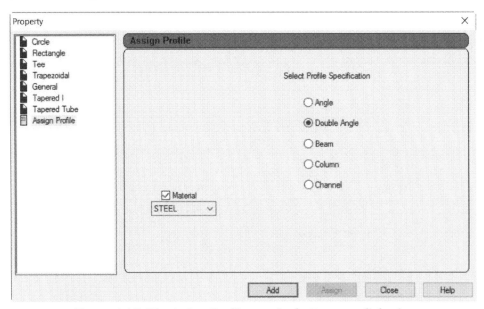

*Figure 4-15 The **Assign Profile** page in the **Property** dialog box*

Steel Sections

STAAD.Pro has in-built steel tables which are used in different countries. It contains thousands of steel sections. To access the steel library, choose the **Section Database** button from the **Properties - Whole Structure** window; the **Section Profile Tables** dialog box will be displayed with the **Steel** tab chosen by default, as shown in Figure 4-16. This dialog box comprises of four tabs: **Steel, Coldformed Steel, Timber,** and **Aluminum**. In the left pane of the **Steel** tab, you can view various sections of different shapes. These sections are grouped into different folders according to their country name. For example, in **American** folder, you can view the American sections. By default, **W Shape** is selected in the left pane and the **W Shape** page is displayed in the right pane of the **Steel** tab. In this page, you can select the required beam from the **Select Beam** list box. You can also specify additional specifications such as Single section, Tee section, Composite section, Top & Bottom Cover plate, and so on in the **Type Specification** area. To include the material constants, select the **Material** check box if it is not selected by default. Next, choose the **Add** button to add the material in the **Material -- Whole Structure** window. Now, you can assign the added material to the structure.

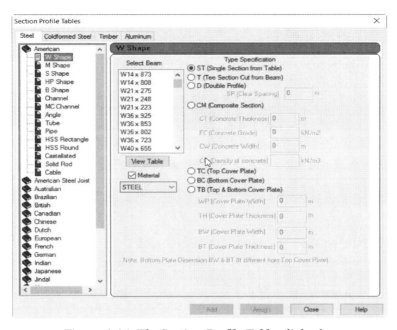

*Figure 4-16 The **Section Profile Tables** dialog box*

Note
*The **American** option is also available in the **Database** drop-down of the **Beam Profiles** panel in the **Specification** tab. On choosing the **American** option; the **American Steel Table** dialog box will be displayed, as shown in Figure 4-17.*

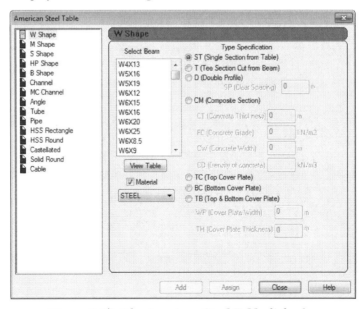

*Figure 4-17 The **American Steel Table** dialog box*

To add the cold formed steel sections to the structure, choose the **Coldformed Steel** tab. In this tab, expand the **American Cold Formed** folder in the left pane if it is not selected by default and select the required shape. Next, select the required section from the right pane of the dialog box. In this pane, specify the type of specification and then choose the **Add** button to add the specification to the **Material - - Whole Structure** window. Next, close the dialog box by choosing the **Close** button. Assign the added section to the structure.

After assigning the section to the structure, you can view the associated commands. To do so, choose the **Editor** button from the **Utilities** panel of the **Utilities** tab; the **Warning** message box will be displayed. Next, choose the **Save** button; the **Editor** window will be displayed. In this window, the command for the assigned steel section is given below the material constants command. The command for the assigned steel section is given next.

```
MEMBER PROPERTY AMERICAN
MEMBER-LIST TABLE ST {CROSS-SECTION TYPE}
```

In the above command, the first line will initiate the command. The Member-List command represents the list of members to which the steel section has been assigned. The Table ST command represents the command for the steel table, and the command in brackets represents the cross section to be assigned.

After assigning the steel section, you can view the rendered view in the **3D Rendering** tool. Figure 4-18 shows the **Rendered View** window with the rendered view of a structure on which the steel section has been applied. You can also view the rendered view of the structure in the main window. To do so, right-click in the main window; a shortcut menu will be displayed. From this menu, choose the **Structure Diagrams** option; the **Diagrams** dialog box will be displayed. In this dialog box, the **Structure** tab is chosen by default. In this tab, select the **Full Sections** radio button in the **3D Sections** area and choose **Apply** and then the **OK** button; the dialog box will be closed and the rendered view of the structure will be displayed, refer to Figure 4-19.

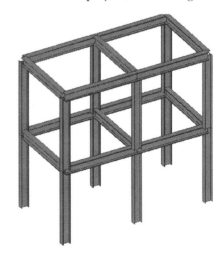

Figure 4-18 Rendered view of the structure in the **Rendered View** window

Figure 4-19 Rendered view of the structure

Steel Joist and Joist Girders

Similar to sections, the steel joist and joist girder sections can also be assigned to a structure. These sections are available in the **Section Profile Tables** dialog box. Expand the **American Steel Joist** node in the left pane of the dialog box; the **K Series** page will be displayed in the right pane. In this page, you can select the required profile from the **Select Profile** list box. After specifying all necessary options, choose the **Add** button to add the section to the **Properties - Whole Structure** window and then close the **Section Profile Tables** dialog box. Now, you can assign the added section to the structure.

You can also find the suitable joist by specifying the values for **Span of Joist**, **Depth of Joist**, **Total Load** and **Live Load** in their corresponding edit boxes in the **Find suitable joist section** area. After specifying the values, choose the **Find Section** button; the appropriate profile will be selected and highlighted in the **Select Profile** list box. Next, choose the **Add** button to add the section and then assign it to the structure.

Plate/Surface Thickness

In STAAD.Pro, you need to specify the thickness for plates and surface elements. To do so, choose the **Thickness** button in the **Properties - Whole Structure** window; the **Plate Element/Surface Property** dialog box will be displayed. In this dialog box, the **Plate Element Thickness** option will be selected in the left pane with the **Plate Element Thickness** page is displayed by default in the right pane, refer to Figure 4-20. In the **Plate Element Thickness** page, you can specify the thickness at each node in the **Node 1**, **Node 2**, **Node 3**, and **Node 4** edit boxes, respectively. For the plates of uniform thickness, specify the value in the **Node 1** edit box and the values in the other edit boxes will be filled automatically. Select the **Material** check box to include the material constants and then select the required material from the drop-down list available below it. Next, choose the **Add** button to add the plate thickness in the **Section** tab. Close the dialog box and then assign thickness to the plates.

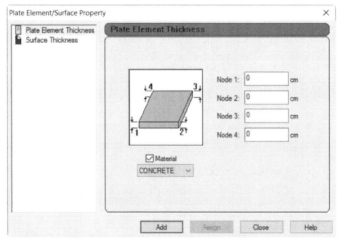

Figure 4-20 *The **Plate Element Thickness** page in the **Plate Element/Surface Property** dialog box*

Similarly, you can add and assign thickness to the surface. To do so, select the **Surface Thickness** option from the left pane of the **Plate Element/Surface Property** dialog box; the **Surface Thickness** page is displayed. In this page, specify the thickness value in the **Node 1** edit box.

In the **Properties - Whole Structure** window, you can choose the **Materials** button to view the values of material constants. On choosing this button, the **Materials** window will be displayed. In this window, the material constants and their properties are listed in a table. Similarly, you can view the section property information by choosing the **Values** button. On doing so, the **Section Properties** table will be displayed. In this table, the added section properties are categorized into tabs. Using the **User Table** button, you can access the user defined sections if they have been defined by the user. If there are no user defined sections, then on choosing this button, the **STAAD.Pro CONNECT Edition** message box will be displayed prompting you to create a section. If you want to create a user defined section, choose the **Yes** button, otherwise choose the **No** button. A cross-section can be removed from the **Section** tab by choosing the **Delete** button. On doing so, the cross-section will be removed from the structure as well. To modify the cross-section properties, choose the **Edit** button.

In the **Beta Angle** tab, you can change the orientation of a member. To do so, choose the **Beta Angle** tab and then choose the **Create Beta Angle** button; the **Beta Angle** dialog box will be displayed, as shown in Figure 4-21. In this dialog box, select the **Angle in Degrees** radio button to specify the angle in degrees. Next, specify the angle in the edit box displayed next to the radio button. Next, choose the **OK** button; the beta angle will be added to the **Beta Angle** tab. Now, select the added angle and then assign it to the required structure by using any of the assignment methods. You can view the oriented members in the rendered view.

*Figure 4-21 The **Beta Angle** dialog box*

Example 3

In this example, you will add and assign steel sections to the structure. After assigning the steel sections, you will orient the member.

Steps to complete this example are given below:

Step 1: Create a new file *c04_staad_connect_ex3* and create a portal frame structure of 5*5m using the **Snap Node/Beam** tool, refer to Figure 4-22.

Step 2: Choose the **Properties** tab and the **Properties - Whole Structure** window is displayed.

Step 3: Choose the **Section Database** button from the **Properties - Whole Structure** window; the **Section Profile Tables** dialog box is displayed.

Step 4: In this dialog box, ensure that the **W Shape** option is selected under the **American** head in the left pane of the dialog box. Select the **W10X26** option from the **Select Beam** area. Ensure that the **Material** check box is selected and the **STEEL** option is selected in the drop-down list below this check box. Next, choose the **Add** button.

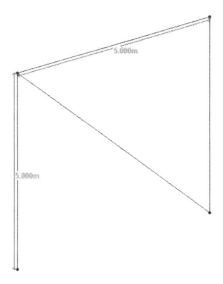

Figure 4-22 The portal frame structure

Step 5: Select the **Angle** option from the left pane of the dialog box; the **Angle** page is displayed in the right pane of the dialog box.

Step 6: Select the **L40406** option from the **Select Angle** area. Next, choose the **Add** button and close the dialog box.

Step 7: In the **Section** tab of the **Properties - Whole Structure** window, select the **W10X26** option. Next, choose the **Beams Parallel To > X** tool from the **Beam cursor** drop-down in the **Selection** panel of the **Geometry** tab; beams parallel to x axis get selected.

Step 8: Ensure that the **Assign To Selected Beams** radio button is selected in the **Assignment Method** area of the **Properties - Whole Structure** window. Choose the **Assign** button; the **STAAD.Pro CONNECT Edition** message box is displayed. Choose the **Yes** button; the selected section is assigned to the selected members.

Step 9: Select the **L40406** option and choose **By Missing Property** tool from the **By Missing Properties** drop-down in the **Selection** panel of the **Geometry** tab; beams are selected.

Step 10: Choose the **Assign** button; the **STAAD.Pro CONNECT Edition** message box is displayed. Choose the **Yes** button; the selected section gets assigned. Choose the **3D Rendered View** button from the **Windows** panel of the **View** tab; the rendered view of the structure is displayed.

Step 11: Choose the **Beta Angle** tab from the **Properties - Whole Structure** window and then choose the **Create Beta Angle** button; the **Beta Angle** dialog box is displayed.

Step 12: Specify **90** in the edit box next to the **Angle in Degrees** radio button and choose the **OK** button.

Step 13: Select the members parallel to x axis and assign **Beta 90** to the selected member using the Assign To Selected Beams method discussed in the previous steps. Figures 4-23 and 4-24 show the structure members before and after orientation.

Step 14: Choose the **Save** button to save the file. Close the file using the **Close** option in the **File** menu.

   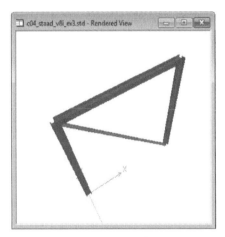

Figure 4-23 *Structure before orientation*   *Figure 4-24* *Structure after orientation*

Self-Evaluation Test

Answer the following questions and then compare them to those given at the end of this chapter:

1. The _____ tool is used to assign material to the selected members.

2. The prismatic sections are defined in the _____ dialog box.

3. In the _____ page, you can define the section properties.

4. You cannot define the tapered sections in STAAD.Pro. (T/F)

5. In STAAD.Pro, you need to specify thickness for the plates. (T/F)

6. In STAAD.Pro, you cannot change the orientation of a member. (T/F)

Review Questions

Answer the following questions:

1. Which of the following buttons is used to invoke the **Section Profile Tables** dialog box?

   (a) **Define**   (b) **Section Database**
   (c) **Edit**   (d) **Delete**

2. Which of the following buttons is used to invoke the **Property** dialog box?

   (a) **Value**   (b) **Define**
   (c) **Materials**   (d) None of these

3. In STAAD.Pro, you cannot create user-defined tables. (T/F)

4. You can edit the property values for the pre-defined materials. (T/F)

5. In isotropic materials, the material properties vary in primary directions. (T/F)

Answers to Self-Evaluation Test

1. Assign To Selected Members, 2. Property, 3. Property, 4. F, 5. T, 6. F

Chapter 5

Specifications and Supports

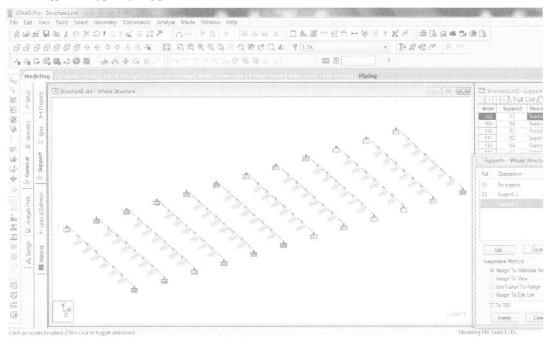

INTRODUCTION

In this chapter, you will learn about various structural conditions and methods of creating and assigning supports to a structure.

In a structural model, you can define the structural conditions of a node, member, and plates. These conditions include specifications such as master/slave nodes, plane stress, fire proofing, member offsets, and so on. In this chapter, these specifications have been categorized under the joint, member, and plate categories. These specifications are discussed next.

NODE SPECIFICATION

In STAAD.Pro, you can assign Master/Slave specification to nodes in a structure. These nodes will act as rigid links in the structure. These structures can be called as rigid diaphragm structures. A rigid diaphragm transfers lateral load to frames or shear walls. A lateral load can be wind load, earthquake load, hydrostatic pressure, and so on. You can define a rigid diaphragm by specifying the rigid links in the structure. In STAAD.Pro, these rigid links are represented as master nodes. To specify a rigid link, choose the **Specification** tab and then select the **Specifications** page from the STAAD.Pro interface; the **Specifications - Whole Structure** window will be displayed in the Data Area. In this window, choose the **Node** button; the **Node Specification** dialog box will be displayed, as shown in Figure 5-1.

Figure 5-1 The **Node Specification** *dialog box*

 In this dialog box, you need to specify the master node which will be the rigid link. The master node is called the master node because the displacements of the slave nodes will be dependent on the displacement of the master node. To specify the master node, select the desired node number from the **Master Node** drop-down list. Next, you need to specify the slave directions. To do so, you need to specify the degrees of freedom with which the slave nodes will be linked to the master node. To make the structure rigid, select the **Rigid** check box. On clearing the **Rigid** check box, the other options in the **Slaved Directions** area will be enabled.

To specify the rigidity limited to in-plane directions, select the **XY**, **YZ**, or **ZX** check box. For example, to specify rigidity in global X and Z axes with rotation about Y axis, select the XY check box. You can also link some specific degrees of freedom by selecting the **FX**, **FY**, **FZ**,

MX, **MY**, or **MZ** check box. After specifying the required conditions, choose the **Add** button; the specification will be added to the **Specifications - Whole Structure** window. Now, you can assign specification to the required node(s) by using any of the assignment methods discussed in Chapter 4. Choose the **Editor** tool from the **Utilities** panel of the **Utilities** tab; the **STAAD Editor** window will be displayed. In the **STAAD Editor** window, you can view the commands for the Master/Slave specification which will be in the format given next.

```
SLAVE {XY, YX, ZX, RIGID, FX, FY, FZ, MX, MY, OR MZ} MASTER i
JOINT n1, n2, n3,......
```

In the above command, the **SLAVE** initiates Master/Slave specification. You can specify the slaved directions by specifying any of the directions enclosed in the bracket. The master node is denoted by letter **i** and the slave nodes are denoted by **n1**, **n2**, **n3**, and so on. You can use the **Delete** button to remove any specification from the **Specification** area. After assigning the master slave specification to the desired node, make sure that the **Highlight Assigned Geometry** check box is selected in the **Specifications - Whole Structure** window. On doing so, master nodes and slave nodes will be highlighted. You can also configure the display of the master/slave nodes. To do so, right-click in the main window to display a shortcut menu. Next, choose the **Labels** option from the menu; the **Diagrams** dialog box will be displayed. In this dialog box, select the **Master Slave (L)** check box in the **General** area. Next, choose the **Apply** and **OK** buttons; the dialog box will be closed and you can see the lines emerging from the master node and linking with the slave nodes. The master node is represented by a cube symbol.

You can create rigid floor diaphragm without specifying master joint by using the **Node Specification** window. Choose the **Node** button; the **Node Specification** window will be displayed. Figure 5-2 shows the **Floor Diaphragm** tab in the **Node Specification** window.

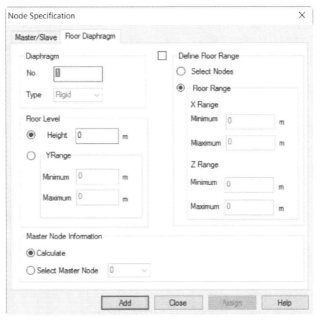

*Figure 5-2   The **Floor Diaphragm** tab in the*
***Node Specification** window*

You can define floor range either by selecting nodes or by providing the range of X and Z coordinates. In both the methods, first ensure that the **Define Floor Range** check box is selected. Now to assign the range using nodes, select the **Select Nodes** radio button from the **Define Floor Range** area. To define floor range by providing range of X and Y coordinates, select the **Floor Range** radio button. As you select the **Floor Range** radio button, the **Minimum** and **Maximum** edit boxes in the **X Range** and **Z Range** areas will become editable. You can enter the required values in the edit boxes in these areas.

After specifying the ranges, select the **Calculate** radio button in the **Master Node Information** area and choose the **Add** button; the specification will be added to the **Specifications -Whole Structure** window and will be assigned to the structure. Choose the **STAAD Editor** button from the **Utilities** panel of the **Utilties** tab; the **STAAD.Pro Editor** window will be displayed.

Note
*The **Floor Diaphragm** and **Master-Slave** commands cannot be specified together.*

In the **STAAD.Pro Editor** window, you can view the commands for the Master/Slave specification which will be in the format given next.

DIA i1 TYPE RIG YR f1 f2JOINT XR f3 f4 ZR f5 f6 ni TO nn

In the above command, the **DIA** command is used for initiating the diaphragm specification.**i1** represents diaphragm identification number. **f1** and **f2** represent Global coordinate values for minimum and maximum Y range respectively. **f3** and **f4** represent Global coordinate values for minimum and maximum X range respectively. **f5** and **f6** represent Global coordinate values for minimum and maximum Z range respectively. **ni** represents node one. **nn** represents nth node in the structure.

Note
In this chapter, you need to download the c04_staad_connect.zip and c05_Staad_connect.zip files for the following examples from http://www.cadcim.com. The path of the file is as follows: Textbook > Civil/GIS > STAAD.Pro > Exploring Bentley STAAD.Pro CONNECT Edition.

Example 1

In this example, you will open the model in *c04_staad_connect_ex2.std* file. Next, you will define the master-slave specification for a rigid diaphragm in which in-plane actions are rigid but bending actions are flexible.

Steps required to complete this example are given next:

Step 1: Open the model *c04_staad_connect_ex2.std* in STAAD.Pro and press SHIFT+N to view the node numbers.

Step 2: Choose the **Specifications** tab and go to the **Specifications** page; the **Specifications - Whole Structure** window is displayed in the data area.

Step 3: Choose the **Node** button; the **Node Specification** dialog box is displayed, refer to Figure 5-3.

Step 4: In this dialog box, select the node number **22** from the **Master Node** drop-down list.

Step 5: Clear the **Rigid** check box in the **Slaved Directions** area and select the **ZX** check box, as shown in Figure 5-3.

*Figure 5-3 The **Node Specification** dialog box*

Step 6: Choose the **Add** button; the **Node Specification** dialog box is closed and the specification is added to the **Specifications - Whole Structure** window.

Step 7: In the **Specifications - Whole Structure** window, ensure that the **SLAVE ZX MASTER 22 JOINT** option is selected. Choose the **front view** tool from the **Tools** panel of the **View** tab; the front view of the structure is displayed.

Step 8: Ensure that the **Nodes Cursor** is chosen from the **Selection** panel of the **Geometry** tab. Click to begin the selection, hold down the left mouse button, and then drag the pointer over the nodes to select; a selection window is created, refer to Figure 5-4.

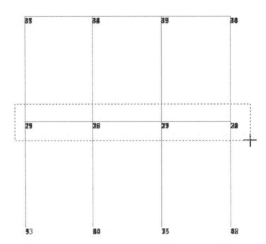

Figure 5-4 *Selecting nodes in the structure*

Step 9: In the **Specifications - Whole Structure** window, ensure that the **Assign To Selected Nodes** radio button is selected and then choose the **Assign** button; the **STAAD.Pro CONNECT Edition** window is displayed. Choose the **Yes** button.

Step 10: Choose the **Isometric View** tool from the **Tools** panel of the **View** tab; the isometric view is displayed. Click in the user interface screen to deselect the nodes.

Step 11: Select the **Highlight Assigned Geometry** check box in the **Specifications - Whole Structure** window; all the slaved nodes as well as the master node is highlighted in the main window. Select the **SLAVE ZX MASTER 22 JOINT** in the specification area; the master node is highlighted with the circles around it.

Step 12: Choose the **Editor** tool from the **Utilities** panel of the **Utilities** tab to view the commands for master-slave specification; the **Warning** message box is displayed prompting you to save the file. Choose the **Save** button; the **STAAD Editor** window is displayed. In this window, you can view the commands for master-slave specification, refer to Figure 5-5. Next, close the **STAAD Editor** window.

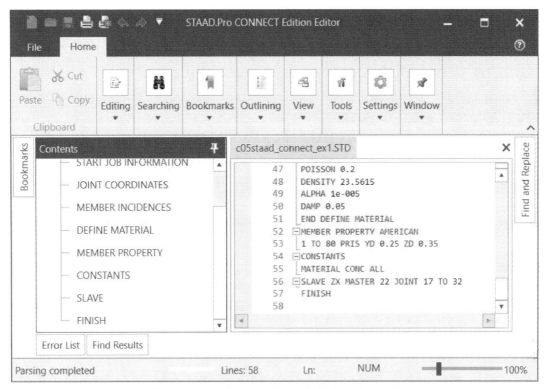

*Figure 5-5 The **STAAD Editor** window displaying the specification syntax*

Step 13: Next, click the right button of the mouse, and select **Labels** option; a dialog box is displayed with the **Labels** tab chosen.

Step 14: Select the **Master Slave (L)** check box in the **General** area of the **Diagrams** dialog box and choose **Apply** and then the **OK** button to close the dialog box; the master/slave nodes are highlighted, refer to Figure 5-6. You can also use the SHIFT+L keys to display master/slave nodes.

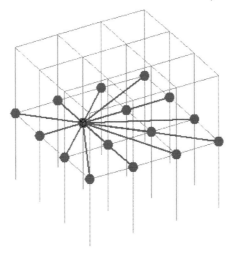

Figure 5-6 The master/slave nodes

Step 15: Choose the **Save As** option from the **File** menu; the **Save As** dialog box is displayed. In this dialog box, specify the name *c05_staad_connect_ex1* in the **File name** edit box and save it at an appropriate location.

Step 16: Close the file by choosing the **Close** option in the **File** menu.

MEMBER SPECIFICATION

Member specifications are used to specify the member conditions such as member release, offset, tension, compression, and so on. These conditions can be specified by using the **Beam** button available in the **Specifications - Whole Structure** window. On choosing this button, the **Member Specification** dialog box will be displayed, as shown in Figure 5-7.

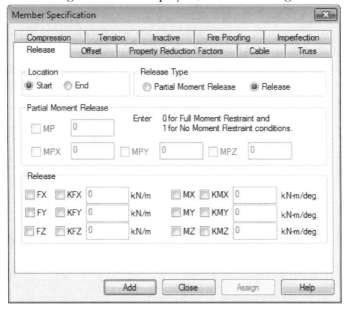

*Figure 5-7 The **Member Specification** dialog box*

This dialog box comprises of following tabs: **Release**, **Offset**, **Property Reduction Factors**, **Cable**, **Truss**, **Compression**, **Tension**, **Inactive**, **Fire Proofing**, and **Imperfection**. Using these tabs, you can define and assign various member conditions. The options in these tabs are discussed next.

Release

In this tab, the release specification is provided in the member end-points to prevent one or more forces/moments to be transferred from one member to the other. You can specify the end conditions for the members by releasing the specified degrees of freedom. To specify the release condition of the member, choose the **Release** tab in the **Member Specification** dialog box, if it is not chosen by default. Various options displayed in this tab are discussed next.

In this tab, select the **Start** or **End** radio button in the **Location** area to define the location of the member end to be released. You can specify the release type by selecting the **Partial Moment Release** or **Release** radio button from the **Release Type** area. If you select the **Partial Moment Release** radio button, then the options in the **Partial Moment Release** area will be enabled.

In this area, you can specify the release factor. You can specify a single partial release factor for MX, MY, and MZ moments. To do so, select the **MP** check box; the edit box next to it will be enabled. Specify the required release factor in this edit box. You can specify the independent factors for the moments by selecting the **MPX**, **MPY**, or **MPZ** check box and then specifying the required factor in the corresponding edit boxes. After specifying the factor, choose the **Add** button; the specification will be added to the **Specifications - Whole Structure** window. Next, assign the specification to the required members by using any of the assignment methods. You can completely release the member end conditions. To do so, select the **Release** radio button. Next, define the release condition by selecting the **FX**, **FY**, **FZ**, **MX**, **MY**, or **MZ** check box. You can also define the spring constants by selecting the **KFX**, **KFY**, **KFZ**, **KMX**, **KMY**, or **KMZ** check box and then specifying the value in the corresponding edit box. Next, choose the **Add** button to add the specification and assign it to the required structure members.

The commands for defining and assigning the release specification is given below:

```
MEMBER RELEASE
MEMBER-LIST {START, END, OR BOTH} {FX, FY, FZ, MX, MY, MZ} {KFX,
KFY, KFZ, KMX, KMY, KMZ}
```

In the above command, the **MEMBER RELEASE** command is used for initiating the release command. The **MEMBER-LIST** command denotes the list of members to be released. The commands for defining the location of the member is enclosed in the brackets next to the **MEMBER-LIST** command. The **FX** to **MZ** commands represent the degrees of freedom to be released and **KFX** to **KMZ** represents the spring constants to be attached.

Offset

To define offset, choose the **Offset** tab in the **Member Specification** dialog box. In this tab, first you need to specify the location on the member, where offset will be provided. To do so, select an option from the drop-down list in the **Location** area. Select the **Start** option to provide an offset on the start point of the member. Select the **End** option to provide an offset on the end point of the member. Next, specify the direction for the offset by selecting an option from the drop-down list in the **Direction** area. The **Global** option is selected by default. To assign the offset to a local axis system, select the **Local** option from the drop-down list. Now, specify the offset distance from the start/end node in the three global directions in the **X**, **Y**, and **Z** edit boxes in the **Offsets** area. Next, choose the **Add** button to add the specification in the **Specifications** area of the **Specifications - Whole Structure** window. Now, select the added specification and assign it to the desired member.

 Note

*After assigning offset to the structural members, you can notice the difference in the members in the **Rendered View** window. Alternatively, you can view the full section in the main window.*

The commands for defining and assigning the offset is given below:

```
MEMBER OFFSET
MEMBER-LIST {START, END} LOCAL {X_d, Y_d, Z_d}
```

In the above command, the **MEMBER OFFSET** command is used for initiating the offset specification. In the **MEMBER-LIST** command, the member numbers are to be specified to which the offset will be applied. Next, the location where the offsets will be applied on the member are specified in the same command. The **LOCAL** command is for specifying the local coordinates for the offset location. If you do not specify the **LOCAL** command then the coordinates will be read in global system.

Example 2

In this example, you will open the model in the *c04_staad_connect_ex3.std* file. Next, you will define and assign offset for the members in a space frame structure.

Steps required to complete this example are given next:

Step 1: Start STAAD.Pro and open the *c04_staad_connect_ex3.std* file; the model is displayed in the main window, refer to Figure 5-8.

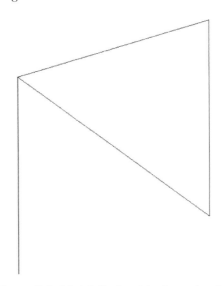

Figure 5-8 Model displayed in the main window

Step 2: Select the **Specifications** page from the **Specification** tab; the **Specifications - Whole Structure** window is displayed in the data area of the interface.

Step 3: In the **Specifications - Whole Structure** window, choose the **Beam** button; the **Member Specification** dialog box is displayed.

Step 4: In this dialog box, choose the **Offset** tab and ensure that the **Start** and **Global** options are selected in the drop-down list in the **Location** and **Direction** areas, respectively.

Step 5: Specify the value **0.073** in the **X** edit box in the **Offsets** area and choose the **Add** button; the dialog box closes and the **START 0.073 0 0** offset is added to the **Specification** area in the **Specifications - Whole Structure** window.

Step 6: Repeat the procedure followed in steps 3 and 4 and select the **End** option from the drop-down list in the **Location** area and the **Global** option from the **Direction** area.

Step 7: Specify the value **-0.073** in the **X** edit box in the **Offsets** area. Next, choose the **Add** button; the dialog box closes and the **END -0.073 0 0** offset is added to the **Specification** area in the **Specifications - Whole Structure** window.

Step 8: Invoke **Beams Cursor** from the **Selection** panel of the **Geometry** tab and click on main window, press **SHIFT+B** and select member **2** from the structure in the main window.

Step 9: Select the **START 0.073 0 0** option from the **Specifications - Whole Structure** window. Next, ensure that the **Assign To Selected Beams** radio button is selected in the **Assignment Method** area and then choose the **Assign** button; the **STAAD.Pro CONNECT Edition** message box is displayed. Choose the **Yes** button; the offset is assigned at the start of the member.

Step 10: Select the **END -0.073 0 0** option from the **Specifications - Whole Structure** and choose the **Assign** button; the **STAAD.ProcConnect Edition** window is displayed. Choose the **Yes** button; the offset is assigned at the end of the member. Choose the **3D Rendering tool** from the **Windows** panel of the **View** tab; the **Rendered View** window is displayed in which you can notice the difference in the members after the offset is assigned, refer to Figure 5-9 and Figure 5-10.

Step 11: Choose the **Save As** option from the **File** menu; the **Save As** dialog box is displayed. In this dialog box, specify the name *c05_staad_connect_ex2* in the **File name** edit box and save it at an appropriate location.

Step 12: Close the file by choosing the **Close** option in the **File** menu.

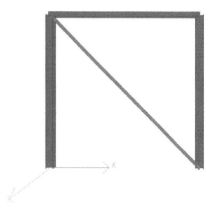

Figure 5-9 *The rendered view before assigning offset*

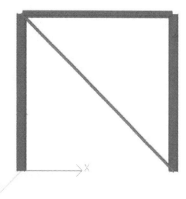

Figure 5-10 *The rendered view after assigning offset*

Property Reduction Factors

In the **Property Reduction Factors** tab, you can specify the factors for reducing the cross-section properties. After reducing the cross-section properties, the structure can be reanalyzed and redesigned. In this tab, you can specify the value for the reduction factors for the cross-sectional area, torsional constant, and moment of inertia in the corresponding edit boxes. After specifying the required factors, choose the **Add** button; the reduction factor specification will be added. Next, assign the reduction factor specification to the required structural member.

Cable

In the **Cable** tab, you can define the cable members. In this tab, you need to specify either the initial tension or the unstressed length of the cable. Specify the initial tension value in the **Initial TENSION** edit box. The specified initial tension is applied as an external load on the structure and it also modifies the stiffness of the member. This tension is used to determine the unstressed length. You can also specify the unstressed length value in the **Unstressed Length** value edit box. This length will be used for the non-linear cable analysis. After specifying the required value, choose the **Add** button to add the specification. Select and assign the specification to the required member.

The commands for defining and assigning a cable is discussed next.

```
MEMBER CABLE
MEMBER-LIST TENSION f₁
```

In the command lines, the **MEMBER CABLE** command is used to start the cable command. In the next line, **MEMBER-LIST** command is used to specify the list of members to be assigned as cables. The **TENSION f_1** command is used to specify the initial tension in cable members.

Truss

In this tab, truss members are subjected to axial loads only. The only degree of freedom for a truss element is axial displacement at each node. To specify the truss members, choose the **Truss** tab and then choose the **Add** button; the **MEMBER TRUSS** specification will be added to the **Specifications** area. Next, select the truss specification and assign it to the required members.

The commands to define and assign the truss members are as follows:

```
MEMBER TRUSS
MEMBER-LIST
```

In the above lines, the **MEMBER TRUSS** command is used to start with the truss command. In the next line, the **MEMBER-LIST** command represents the list of members that will be assigned as truss members.

Compression

The compression members carry compressive forces only and can be specified using this tab. To specify the compression members, choose the **Compression** tab and then choose the **Add** button; the **MEMBER COMPRESSION** specification will be added to the **Specifications** area. Next, select it and assign it to the required members.

Tension

The tension members carry tensile forces only and can be specified using this tab. To specify the tension members, choose the **Tension** tab and then choose the **Add** button; the **MEMBER TENSION** specification will be added to the **Specifications** area. Next, select the added specification and assign it to the required members.

Note

*The commands for specifying tension and compression members is **MEMBER TENSION** and **MEMBER COMPRESSION**, respectively. Similarly, the command **MEMBER TRUSS** is for specifying truss members.*

Inactive

Using this tab, you can make some members inactive while analyzing a structure. The stiffness contribution of these inactive members will not be considered during the analysis. To make the members inactive, choose the **Inactive** tab and then choose the **Add** button; the **Inactive Member** specification will be added to the **Specifications** area. Next, select the added specification and assign it to the required members.

The command for defining and assigning inactive specification is as follows:

```
INACTIVE MEMBER-LIST
```

You can make the members active again by using the **CHANGE** command. To do so, choose the **Analysis Commands** option in the **Analysis Data** panel of the **Analysis & Design** tab; the **Analysis/Print Commands** dialog box will be dispalyed. In this dialog box, choose the **Change** tab and then choose the **Add Button** button. Next, close the dialog box. You can specify the **CHANGE** command only after the **PERFORM ANALYSIS** command.

Fire Proofing

In STAAD.Pro, you can calculate the weight of fire proofing material applied to a structure. To do so, choose the **Fire Proofing** tab. In this tab, you can specify two types of fire configurations: Block Fire Proofing and Contour Fire Proofing. In block fire proofing, a rectangular block of fire proofing material will be formed around the steel section. To define block fire proofing, select the **BFP (Block Fire Proofing)** radio button and specify the thickness and density of material in their respective edit boxes.

In contour fire proofing, a coating of fire proofing material will be formed around the steel section. To define contour fire proofing, select the **CFP (Contour Fire Proofing)** radio button and then specify the thickness and density of material in their respective edit boxes.

After specifying the required fire proofing type, choose the **Add** button; the specification will be added to the **Specifications** area of the **Specifications - Whole Structure** window. Next, select the added specification and assign it to the required structure member.

Imperfection

The specifications of the drift and cambers can be defined in the **Imperfection** tab. In this tab, select the **Camber** radio button. On doing so, the related parameters will be displayed in this tab. Next, specify the parameters for the camber and choose the **Add** button to add the camber specification. Next, select the added specification and assign it to the required members.

Similarly to define drift, select the **Drift** radio button and then specify the required parameters and choose the **Add** button to add the drift specification. Next, select the added specification and assign it to the required members.

PLATE SPECIFICATION

The plate specification includes Element Releases, Ignore Inplane Rotation, Rigid Inplane Rotation, Plane Stress, and Ignore Stiffness. These specifications can be defined by choosing the **Plate** button available in the **Specifications - Whole Structure** window. On choosing the **Plate** button, the **Plate Specs** dialog box will be displayed, as shown in Figure 5-11.

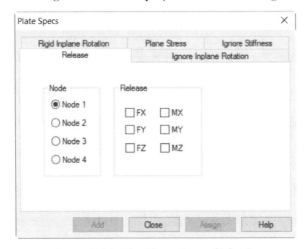

Figure 5-11 The **Plate Specs** *dialog box*

This dialog box comprises of five tabs: **Release**, **Ignore Inplane Rotation**, **Rigid Inplane Rotation**, **Plane Stress**, and **Ignore Stiffness**. Using these tabs, you can define the plate specifications. These tabs are discussed next.

Release

Using the options in the **Release** tab, you can release one or more degrees of freedom at the corner nodes of the plate element. To specify the releases to a node, select the **Node 1**, **Node 2**, **Node3**, or **Node 4** radio button from the **Node** area of the **Plate Specs** dialog box. Next, define the release condition for the translational and rotational degrees of freedom by selecting the corresponding check boxes from the **Release** area. Next, choose the **Add** button to add the release specification and then assign it to the required element.

The commands to define and assign the element release specification is as follows:

```
ELEMENT RELEASE
ELEMENT-LIST {J_1, J_2, J_3, J_4} {FX, FY, FZ, MX, MY, MZ}
```

In the above lines, the **ELEMENT RELEASE** command is used for initializing the release specification. In the next line, the **ELEMENT-LIST** command represents the list of element numbers to which the release specification will be assigned. The $\{J_1, J_2, J_3, J_4\}$ command represent the nodes to be released and the $\{FX, FY, FZ, MX, MY, MZ\}$ command represent the degrees of freedom to be released.

Ignore Inplane Rotation

In this tab, you can define the specification to ignore the in-plane rotation actions. To do so, in the **Ignore Inplane Rotation** tab, choose the **Add** button; the **ELEMENT IGNORE INPLANE ROTATION** specification will be added to the **Specifications** area. Next, select and assign the added specification to the required plate element. The command for applying this specification is as follows:

```
ELEMENT IGNORE INPLANE ROTATION
ELEMENT-LIST
```

Rigid Inplane Rotation

In this tab, you can define the specification to rigid the in-plane rotation actions. To do so, in the **Rigid Inplane Rotation** tab, choose the **Add** button; the **ELEMENT RIGID INPLANE ROTATION** specification will be added to the **Specifications** area. Next, select and assign the added specification to the required plate element. The command for applying this specification is as follows:

```
ELEMENT RIGID INPLANE ROTATION
ELEMENT-LIST
```

Plane Stress

The Plane Stress specification is used to model the selected elements for plane stress only. In this case, no bending or transverse shear is applied. To do so, in the **Plane Stress** tab, choose the **Add** button; the **ELEMENT PLANE STRESS** specification will be added in the **Specifications - Whole Structure** window. Assign the added specification to the plate elements. The command for the Plane Stress specification is given next:

```
ELEMENT PLANE STRESS
ELEMENT-LIST
```

Ignore Stiffness

While modeling the plate elements, you might not want to consider the stiffness of some of the plate elements in the analysis. These elements will carry the loads and transfer them to the other parts of the structure but will not provide any additional stiffness to the structure. In this case, you can use the Ignore Stiffness specification. To do so, in the **Ignore Stiffness** tab, choose the **Add** button; the **IGNORE STIFFNESS ELEMENT** specification will be added to the **Specifications** area of the **Specifications - Whole Structure** window. Assign the added specification to the required plate elements. The command for assigning the Ignore Stiffness specification is as follows:

```
IGNORE STIFFNESS ELEMENT
ELEMENT-LIST
```

SUPPORTS

Generally three types of supports are used to join a structure to its foundation: Fixed, Pinned, and Roller supports. The fourth type is simple support which is not often found in structures. All these supports can be placed anywhere along a structure. For example, these supports can be provided at ends, midpoints, and intermediate points. The support type provided to a structure will determine the type of load a support can resist.

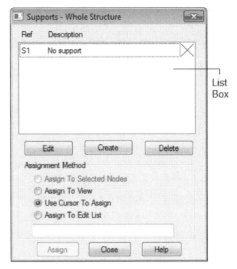

In STAAD.Pro, you can provide different types of supports such as fixed, pinned, roller, inclined, springs, and so on. To define supports, choose the **Supports** page from the use interface screen; the **Supports - Whole Structure** window will be displayed, as shown in Figure 5-12. In this window, choose the **Create** button; the **Create Support** dialog box will be displayed, as shown in Figure 5-13.

*Figure 5-12 The **Supports - Whole Structure** window*

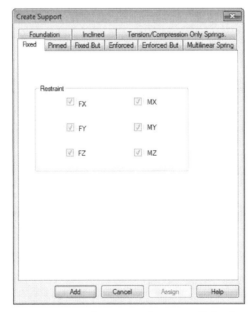

*Figure 5-13 The **Create Support** dialog box*

In this dialog box, all the supports are available in different tabs. These tabs are discussed next.

Fixed

Using this tab, you can define fixed supports. Fixed supports restricts all six degrees of freedom of an element. Along with the moment, they can resist vertical and horizontal forces. So, you can also call them as rigid supports. You will notice that in the **Fixed** tab, all the check boxes for the translational and rotational degrees of freedom are selected by default. You cannot clear

these check boxes. To define fixed supports, in the **Fixed** tab, choose the **Add** button; the
fixed support will be added to the list box with the name **Support 2**. The fixed support
will be represented by a symbol. Next, select the added support and assign it to the appropriate
nodes. You can use any method to assign the supports. These methods are already discussed in
the previous chapter. After assigning the support, choose the **Editor** tool from the **Utilities**
panel of the **Utilities** tab; the **STAAD Editor** window will be displayed. In this window, you can
view the command for assigning the fixed supports.

For assigning supports using the **Editor** tool, first you need to specify the **SUPPORTS** command.
In the next line, you will specify the member numbers to which supports will be assigned and
then specify the type of support to be assigned. The command for assigning the fixed support
is given below:

```
SUPPORTS
MEMBER-LIST FIXED
```

> **Note**
> *The commands for the pinned and enforced supports will be same as for the fixed supports. In this*
> *case, after specifying the member list, you need to specify the type of support. For example, for the*
> *pinned support, the command will be **MEMBER-LIST PINNED**.*

Pinned

Using this tab, you can define pinned supports. Pinned supports restricts the three translational
degrees of freedom of an element, but the three rotational degrees of freedom are set free.
Pinned supports therefore, allows the structural member to rotate but no translational movement
will be possible. To define pinned support, choose the **Pinned** tab. In the **Restraint** area of this
tab, you will notice that the check boxes for the translational degrees of freedom are selected
and for the rotational degrees of freedom are cleared. Choose the **Add** button in this tab; the
pinned support will be added to the list box in the **Supports - Whole Structure** window.
Select the added pinned support and assign it to the appropriate nodes.

Fixed But

Using the options in this tab, you can create roller, hinge, and spring supports with specified
degrees of freedom for an element. Roller supports are free to rotate and move along the surface
on which the support rests. Hinge supports allows rotation only. To define any of the supports
mentioned above, choose the **Fixed But** tab; the release check boxes and the define spring
options will be displayed. To release any of the six degrees of freedom, select the corresponding
check box from the **Release** area of this tab. To define spring support, specify the required value
in the appropriate edit box in the **Define Spring** area. Next, choose the **Add** button to add
the defined support to the list box in the **Supports - Whole Structure** window. Next, select the
added support and assign it to the appropriate nodes. The general format for the command to
assign the **Fixed But** support is given next.

```
SUPPORTS
MEMBER-LIST FIXED BUT {RELEASE-SPECIFICATION}{SPRING CONSTANTS}
```

In the preceeding lines, after specifying the **MEMBER-LIST** and **FIXED BUT** commands, you need to specify the degrees of freedom to be released such as FX, FY, FZ, and so on. You can define springs by specifying the spring specification such as KFX, KFY, KFZ, and so on and the constants value next to it. For example, the command for spring constant can be written as KFY 50.

Enforced

Using this tab, you can define enforced supports. Enforced supports also restricts all degrees of freedom of an element alike the fixed supports. They are used while considering support displacement loads in case of plates and solids. Support displacement loads are not allowed if fixed supports are assigned to the plates and solids. To define enforced supports, choose the **Enforced** tab in the **Create Support** window and choose the **Add** button; the support will be added to the list box in the **Supports - Whole Structure** window. Next, select and assign it to the appropriate nodes.

Enforced But

Using the options in this tab, you can define enforced supports with of the required degrees of freedom set free. To define this support, choose the **Enforced But** tab in the **Create Support** window. To release any of the six degrees of freedom, select the corresponding check box from the **Release** area of this tab. Next, choose the **Add** button; the support will be added in the list box in the **Supports - Whole Structure** window. Next, select and assign it to the appropriate nodes. The command to assign the **Enforced But** support is given next.

```
SUPPORTS
MEMBER-LIST ENFORCED BUT {FX,FY,FZ,MX,MY,MZ}
```

Multilinear Spring

Multilinear spring supports are used while applying static load to a structure. They can be used for defining soil spring supports where the behavior in tension zone differs from the behavior in compression zone. To define this support, choose the **Multilinear Spring** tab. Next, in this tab, specify the values for the displacement of support node and the spring stiffness in the **Displacement** and **Spring Stiffness** columns. Next, choose the **Add** button to add the support. Now, select it and assign at the appropriate nodes.

Foundation

You can define spring supports for footings and mat foundation by using the parameters in the **Foundation** tab. To define a spring support for isolated footing, select the **Footing** radio button in the **Foundation** tab. Specify the dimension of the footing in the **L** and **W** edit boxes. Next, specify the direction of spring by selecting the required radio button in the **Direction** area. In this area, if you select the **X**, **Y**, or **Z** radio button, then the spring will be generated in that direction only and the other degrees of freedom will receive fixed support. If you select **X Only**, **Y Only**, or **Z Only** radio button, then the spring support will be generated in the respective direction only and the rest of the degrees of freedom will be free to deform. You can specify the soil subgrade value in the **Subgrade** edit box in the **Subgrade** area.

You can define spring supports for mat foundations by using two methods: Elastic Mat and Plate Mat methods. In both of these methods, STAAD.Pro will calculate the influence area of the nodes that define the surface. Then this influence area will be multiplied with the soil subgrade modulus to calculate spring stiffness value. In the elastic mat method, the influence area will be calculated for the joints, and in the plate mat method, the influence area will be calculated for the plates. To use elastic mat method, select the **Elastic Mat** radio button in the **Foundation** area. Next, specify the direction and soil subgrade values. To use the plate mat method, select the **Plate Mat** radio button in the **Foundation** area and specify the required parameters.

After specifying the desired spring support parameters, choose the **Add** button to add it to the list box of the **Supports - Whole Structure** window. Next, select and assign it to the appropriate nodes.

Inclined

Using the options in this tab, you can define supports that are inclined with respect to the global axis system. To define an inclined support, choose the **Inclined** tab. In this tab, you can define inclined supports in three ways: By specifying the coordinates of the datum, by specifying the coordinates of the reference point which describes the inclined axis system, and by specifying the reference joint of the support.

In the **Inclined** tab, select the **Coordinate** radio button in the **Incline Reference Point** area to specify the coordinates of the datum of the inclined axis system. Next, specify the coordinates in the **X**, **Y**, and **Z** edit boxes. To define the inclined support by specifying the reference point, select the **Ref** radio button and then specify the coordinates in the **X**, **Y**, and **Z** edit boxes. To define the inclined support by specifying the reference joint, select the **RefJt** radio button and then select the joint number from the drop-down list displayed next to the radio button. Next, you can specify the type of support, release conditions, and spring constant in their respective areas. Next, choose the **Add** button; the support will be added to the list box in the **Supports - Whole Structure** window. Next, select and assign it at the appropriate nodes.

The command for assigning the inclined support by specifying the x, y, and z coordinates of the datum is as follows:

```
SUPPORTS
MEMBER-LIST INCLINE x y z{SUPPORT TYPE}{RELEASE SPEC}{SPRING
CONSTANT}
```

Next, you can specify the support type, release specifications, and spring constants, if required.

The command for assigning the inclined support by specifying the coordinates of the point is as follows:

```
SUPPORT
MEMBER-LIST INCLINE REF x y z{SUPPORT TYPE}{RELEASE SPEC}{SPRING
CONSTANT}
```

The command for assigning the inclined support by specifying the reference joint is as follows:

```
SUPPORTS
MEMBER-LIST INCLINE REFJT n {SUPPORT TYPE}{RELEASE SPEC}{SPRING
CONSTANT}
```

Tension/Compression Only Springs

Using this support type, you can define spring supports as tension only or compression only. To define this support, choose the **Tension/Compression Only Springs** tab. Next, specify the degree of freedom by selecting the **Tension Only** or **Compression Only** radio button. Now, specify the degree of freedom which will be set unidirectional by selecting the check boxes in the **Spring Direction** area. Choose the **Add** button; the support will be added to the list box in the **Supports - Whole Structure** window. Next, select and assign it to the appropriate nodes.

The command for assigning the **Tension/Compression Only Springs** support is given as follows:

```
SUPPORTS
SPRING TENSION/COMPRESSION
JOINT-LIST SPRING SPECIFICATION
```

Example 3

In this example, you will create an inclined portal frame structure and define inclined supports for an inclined axis system.

Steps required to complete this example are given below:

Step 1: Start STAAD.Pro CONNECT Edition and choose the **New** option from the left pane of the user interface screen; the **Model Information** page is displayed. In this page, specify the name *c05_staad_connect_ex3* in the **File Name** edit box and browse to the location *C:\STAAD Examples\c05_staad_connect* by clicking the button next to the **Location** edit box.

Step 2: Select the **Type** as **Analytical** and **Units** as **Metric**. Choose the Create button from the left pane of the user interface screen; the file is loaded. Open the **STAAD Editor** window from the **Utilities** panel of the **Utilities** tab; the **STAAD.Pro Editor** window is displayed.

Step 3: In this window, specify the commands, as shown in Figure 5-14.

Step 4: Choose the **Save** button from the **File** menu in the **STAAD.Pro Editor** window and close it. Press SHIFT+N and SHIFT+B to view the node and beam number. Figure 5-15 shows the model displayed in the main window.

Step 5: Invoke the **Supports** page; the **Supports - Whole Structure** window is displayed in the right area of the interface.

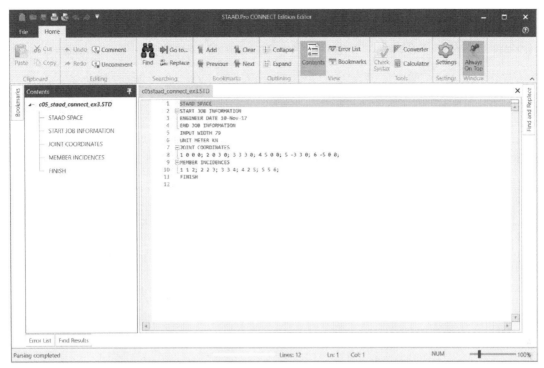

Figure 5-14 *Commands specified in the* **STAAD Editor** *window*

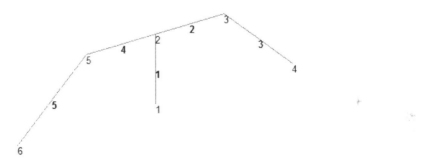

Figure 5-15 *Model displayed in the Main Window*

Step 6: In this window, choose the **Create** button from the **Supports - Whole Structure** window; the **Create Support** dialog box is displayed with the **Fixed** tab chosen.

Step 7: Choose the **Add** button; the **Support 2** is added to the list box in the **Supports - Whole Structure** window. Select the added support and assign it to the node **1** using the **Use Cursor To Assign** method.

Step 8: Invoke the **Create Support** dialog box by choosing the **Create** button and then choose the **Inclined** tab. In this tab, select the **Ref** radio button in the **Incline Reference Point** area.

Step 9: Next, specify the values **3**, **3**, and **0** in the **X**, **Y**, and **Z** edit boxes, respectively and then select the **Fixed** radio button in the **Support Type** area. Next, choose the **Add** button; the **Support 3** is added to the list box in the **Supports - Whole Structure** window.

Step 10: Now, select the added support and assign it to the node number **4**.

Step 11: Again, invoke the **Create Support** dialog box by choosing the **Create** button and then choose the **Inclined** tab. In this tab, select the **RefJt** radio button in the **Incline Reference Point** area.

Step 12: Select the node number **5** in the **Node Number** drop-down list and select the **Fixed** radio button in the **Support Type** area. Next, choose the **Add** button; the **Support 4** is added to the list box in the **Supports - Whole Structure** window.

Step 13: Now, select the added support and assign it to the node number **6**. Figure 5-16 shows the model after assigning the supports.

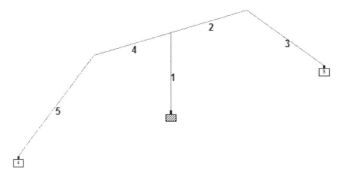

Figure 5-16 *Supports assigned to the structure*

Step 14: Choose the **Editor** tool from the **Utilities** panel of the **Utilities** tab; the **STAAD Editor** window is displayed. In this window, you can view the commands for the added supports as given below:

```
SUPPORTS
1 FIXED
4 INCLINE REF 3 3 0 FIXED
6 INCLINE REFJT 5 FIXED
```

Step 15: Close the **STAAD Editor** window.

 Note
*While closing the **STAAD Editor** window if a warning message box is displayed prompting you to save the file, choose the **Save** button.*

Step 16: Choose the **Save As** option from the **File** menu; the **Save As** dialog box is displayed. In this dialog box, specify the name *c05_staad_connect_ex3* in the **File name** edit box and save it at an appropriate location.

Step 17: Close the file by choosing the **Close** option in the **File** menu.

Self-Evaluation Test

Answer the following questions and compare them to those given at the end of this chapter:

1. Rigid diaphragms are modeled using the _____ specification.

2. The _____ specification is used to release the degrees of freedom at the member ends.

3. The truss members carry _____ loads only.

4. In the _____ tab, you can specify the drift and camber specifications.

5. In cable members, the specified initial tension acts as an external load applied on the structure. (T/F)

6. The Inactive specification is used to make the members inactive. (T/F)

7. The compression members carry tensile forces only. (T/F)

Review Questions

Answer the following questions:

1. Which of the following is used as a type of Fire Proofing specification?

    (a) Block Fire Proofing   (b) Release
    (c) Cable   (d) Both a & b

2. Which of the following specifications is used after the Inactive command to make the members active again?

    (a) Offset   (b) Change
    (c) Inactive   (d) Release

3. Which of the following support types is used to specify supports for static load cases?

    (a) Fixed   (b) Multilinear Spring
    (c) Enforced   (d) Pinned

4. The **Fixed But** tab is used to create roller, spring, and hinge supports. (T/F)

5. Enforced supports are used for support displacement loads. (T/F)

6. The Ignore Stiffness specification is used to ignore the stiffness of the plate elements during analysis. (T/F)

7. Fixed supports are restrained in three translational degrees of freedom only. (T/F)

Answers to Self-Evaluation Test
1. Node, **2.** Release, **3.** axial, **4.** Imperfection, **5.** F, **6.** T, **7.** F

Chapter **6**

Loads

Learning Objectives

After completing this chapter, you will be able to:

- *Define and assign Primary Loads*
- *Define and assign Load Generation*
- *Define and assign Load Combinations*
- *Define and assign Load Generation Automatically*

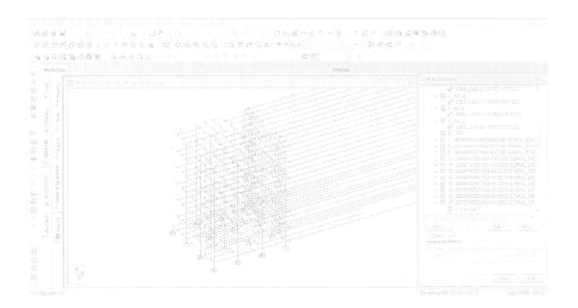

INTRODUCTION

In structural engineering, loads can be classified into different categories: dead loads, live loads, environmental loads, combination loads, and other loads. Dead loads are the loads which remain constant over certain extent of time. It includes weight of the structure such as weight of walls, beams, columns, and so on. Live loads are the moving loads which act for a short duration, for example, a moving vehicle. Environmental loads are the loads which act due to weather and other natural phenomena such as wind, snow, earthquake, and so on. Combination loads are experienced when different types of loads act together. Other loads include support displacement loads, hydrostatic loads, and so on.

Once you model a structure in STAAD.Pro including its cross-sections, supports, and specifications, you can define and assign loads to the structural members. In STAAD.Pro, the above discussed categories are further classified into different groups. These groups are: Primary Loads, Load Generation, Combination Loads, and Auto Load Combination, which are discussed next.

PRIMARY LOADS

In STAAD.Pro, the Primary Loads group includes different types of loads such as Selfweight, Nodal Load, Member Load, and so on. To define these load types, first create the load cases and then add required loads in these load cases.

To do so, choose the **Loading** tab; the **Load & Definition** window will be displayed in the user interface screen, refer to Figure 6-1. Select the **Load Cases Details** node in the list box, refer to Figure 6-1. Next, choose the **Add** button; the **Add New : Load Cases** dialog box will be displayed, as shown in Figure 6-2. In this dialog box, the **Primary** node will be highlighted in the left pane and the **Primary** page will be displayed in the right pane. In the **Primary** page, specify the load case number in the **Number** edit box. Select the type of load such as dead, live, wind, and so on from the **Loading Type** drop-down list. You can also enter load case title in the **Title** text box for your reference. Choose the **Add** button to add the load case and close the dialog box. You can see that the load case will be added to the list box of the **Load & Definition** window. Similarly, you can add more load cases for different loads.

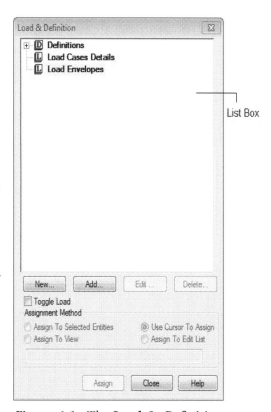

List Box

Figure 6-1 *The* *Load* & *Definition* *window*

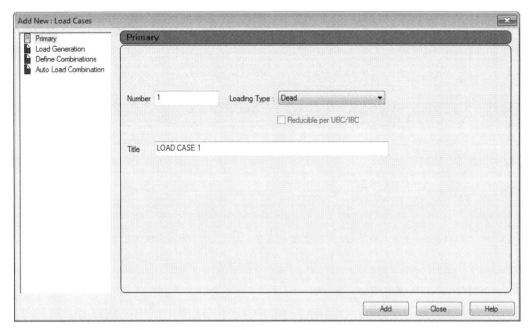

*Figure 6-2 The **Add New : Load Cases** dialog box*

Next, in the added load case, you will define different load types and assign them to the structure. To define these load types, select the required load case from the **Load & Definition** window and choose the **Add** button; the **Add New : Load Items** dialog box will be displayed, as shown in Figure 6-3. Different types of loads available in this dialog box are discussed next in detail.

Selfweight

The selfweight of a structure consists of a major portion of the dead load in it. To add the selfweight of a structure for analysis, expand the **Selfweight** node of the **Add New : Load Items** dialog box; the **Selfweight Load** option will be highlighted and the **Selfweight Load** page will be displayed in the right pane. In this page, specify the direction in which the selfweight will be applied by selecting the required radio button from the **Direction** area. Next, specify the factor value in the **Factor** edit box. This value will be multiplied with the selfweight. A -ve sign against the value represents that the load will be applied along the negative direction of the selected axis. Next, choose the **Add** button; selfweight will be added to the list box in the **Load & Definition** window. A question mark next to the load added in the **Load & Definition** window indicates the unassigned load. Next, you can assign the added selfweight to a structure using the **Assign To View** method which is discussed in the previous chapters. In STAAD.Pro, selfweight of every structural member will be calculated and applied as uniformly distributed load on the members. The command for assigning the selfweight is given next:

```
SELFWEIGHT {X,Y,OR Z} f MEMBER-LIST
```

In the above command, **X**, **Y**, and **Z** represent the global direction in which the selfweight will act and **f** represents the factor whose value is multiplied with the selfweight.

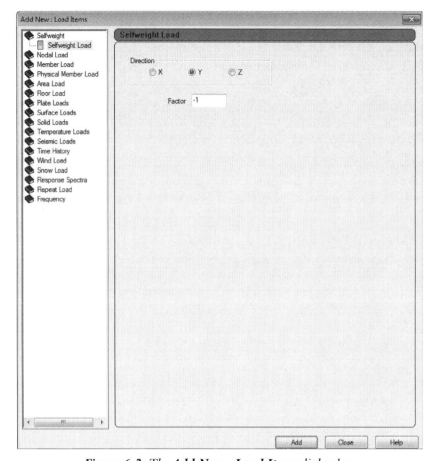

Figure 6-3 The **Add New : Load Items** *dialog box*

Nodal Load

Nodal load is used to define and assign the concentrated loads and moments at the joints of a structure. These loads act in the global coordinate system. To define nodal load, expand the **Nodal Load** node in the left pane of the **Add New : Load Items** dialog box; the **Node** and **Support Displacement** load options will be displayed in the left pane and the **Node** page will be displayed in the right pane of the dialog box. In the **Node** page, specify the values for the forces in the **Fx**, **Fy**, and **Fz** edit boxes. Next, you can specify the values for the concentrated moment in the **Mx**, **My**, and **Mz** edit boxes. After specifying the values, choose the **Add** button; the specified nodal load will be added to the list box in the **Load & Definition** window. A question mark symbol can be seen next to the added load and indicates the unassigned load. Now, select the added load and assign it to the appropriate nodes in the structure using the **Use Cursor To Assign** method, as discussed earlier.

Similarly, you can define and assign support displacement loads. To do so, select the **Support Displacement** load option in the left pane of the **Add New : Load Items** dialog box; the **Support Displacement** page will be displayed in the right pane of the dialog box. In this page, specify the displacement value in the **Displacement** edit box. Specify the direction of displacement such as translational or rotational by selecting the required radio button in the **Direction** area. Next, choose the **Add** button to add the load and then assign it at the appropriate nodes.

Note
*After assigning the load, you will notice that the question mark displayed next to the added load in the **Load & Definition** window gets replaced by a green cube which indicates that the defined load has been assigned to the structure.*

Member Load
Member load is directly applied to the structural member. It includes uniformly distributed loads, concentrated loads, linear varying loads, and so on. Uniform and Linear Varying loads act over the full or partial length of the member. While concentrated loads act at a point on the member. Various types of member loads are discussed next.

Uniform Force and Uniform Moment
Expand the **Member Load** node in the left pane of the **Add New : Load Items** dialog box; the **Uniform Force** and **Uniform Moment** options will be displayed. The **Uniform Force** option will be selected by default in the left pane and the **Uniform Force** page will be displayed in the right pane of the dialog box. In this page, specify the load value in the **W1** edit box in the **Force** area. Next, specify the distance between the start point of the member and load in the **d1** edit box. Similarly, specify the distance between the start point of the member and the end point of the load in the **d2** edit box and the perpendicular distance between member's shear centre to the plane of loading in the **d3** edit box. You can specify the direction of loads by selecting the corresponding radio buttons in the **Direction** area. You can specify direction in the local coordinates by selecting the **X (Local)**, **Y (Local)**, or **Z (Local)** radio button. You can also specify the direction in global coordinates by selecting the **GX**, **GY**, or **GZ** radio button. The **PX**, **PY**, or **PZ** radio button can be selected to define loads along the projected length of the member along the global direction. Next, choose the **Add** button to add the load in the **Load & Definition** list box. Select the added load and assign it to the structural members.

To define uniform moment, select the **Uniform Moment** option under the **Member Load** node in the left pane of the dialog box; the **Uniform Moment** page will be displayed in the right pane. In this page, specify the value of moment in the **W1** edit box in the **Moment** area. Specify the values in the **d1**, **d2**, and **d3** edit boxes. Specify the direction of moment by selecting the corresponding radio buttons in the **Direction** area. Choose the **Add** button to add the moment in the **Load Definition** window. Next, assign the moment to the structural members. Figure 6-4 shows the structure on which uniform load and moment is applied. The command used for assigning uniform force and moment is given next:

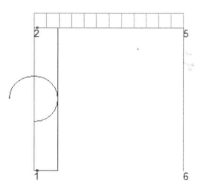

Figure 6-4 Uniform load and moment applied on the member

```
MEMBER LOAD
MEMBER-LIST UNI/UMOM DIRECTION-SPEC {W1} {d1,d2,d3}
```

In the command, **UNI** is used for the uniform force and **UMOM** for the uniform moment. The **DIRECTION-SPEC** command is used for specifying the direction. Here, you need to specify the direction such as **GX**, **GY**, and so on. **W1** represents the command for specifying the force or moment value. **d1** and **d2** represent the distance from the start node of the member to the start and end point of the load. And **d3** represents the placement of load in reference to the center line of the member along the cross-section.

Concentrated Force and Concentrated Moment

Concentrated load may act at any point on the structural member. Figure 6-5 shows the concentrated force and moment applied on a structural member. To define concentrated force, select the **Concentrated Force** option from the **Member Load** node in the left pane of the dialog box; the **Concentrated Force** page will be displayed in the right pane of the dialog box. In this page, specify the value of load in the **P** edit box in the **Force** area. Next, specify the distance of load from the start point of the member in the **d1** edit box. Specify the distance from the member's shear centre to the plane of loading in the **d2** edit box. In the **Direction** area, specify the direction in which load will be applied on the member by selecting the desired radio button. Next, choose the **Add** button to add the load and then assign it at an appropriate place in the structure.

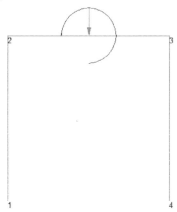

Figure 6-5 *Concentrated force and moment applied on the member*

In the same way, you can define and assign the concentrated moment to a structure. To do so, select the **Concentrated Moment** option under the **Member Load** node; the **Concentrated Moment** page will be displayed. In this page, specify the required values in the same way as discussed earlier for the concentrated force load. The command used for assigning concentrated force and moment is given next:

```
MEMBER LOAD
MEMBER-LIST CON/CMOM DIRECTION-SPEC {W1} {d1,d2,d3}
```

Linear Varying

Linear varying loads act on the whole structural member or on some portion. It acts in a non uniform manner. Figure 6-6 shows the linear varying loads applied on the structural members. To define this load, select the **Linear Varying** load option under the **Member Load** node in the left pane of the **Add New : Load Items** dialog box; the **Linear Varying** page will be displayed in the right pane. In this page, define load in increasing or decreasing manner by specifying the load values in the **W1** and **W2** edit boxes in the **Force** area. In this page, you can also define triangular load. To do so, select the **W3** radio button and specify the load value in the corresponding edit box. Next, specify the direction of loading by selecting the **X (Local)**, **Y (Local)**, or **Z (Local)** radio button. Next, choose the **Add** button to add the load to the **Load & Definition** window and then assign it on the appropriate members. The command used for assigning linear varying load is given next.

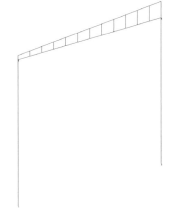

Figure 6-6 *Linear Varying load applied on the structural member*

```
MEMBER LOAD
MEMBER-LIST LIN DIRECTION-SPEC {W1, W2, W3}
```

In the command, **W1** and **W2** represent load values for trapezoidal load and **W3** represents load value for triangular load.

Trapezoidal

Trapezoidal loads are applied in a trapezoidal manner on the structure member, refer to Figure 6-7. To define trapezoidal load, select the **Trapezoidal** load option available under the **Member Load** node; the **Trapezoidal** page will be displayed. In the **Force** area of this dialog box, specify the value of load intensity in the **W1** and **W2** edit boxes. Next, specify the distance of loading from the start point of the member to the start point of loading in the **d1** edit box, and from the start point of the member to the end point of loading in the **d2** edit box. Next, in the **Direction** area, specify the direction of loading by selecting the required radio buttons. Next, choose the **Add** button to add the load and then assign it to the desired structural members. The command used for assigning trapezoidal load is given next.

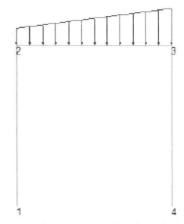

Figure 6-7 *Trapezoidal load applied on the structural member*

```
MEMBER LOAD
MEMBER-LIST TRAP DIRECTION-SPEC {W1, W2} {W3}
```

Hydrostatic

Hydrostatic loads are the loads produced due to hydrostatic pressure. Figure 6-8 shows the hydrostatic load applied on the structural member. To define hydrostatic loads, select the **Hydrostatic** load option under the **Member Load** node; the **Hydrostatic** page will be displayed. In this page, the displayed options will be in an inactive mode. To enable these options, you need to select the member on which the load is to be applied. To enable these options, choose the **Select Member** button; the **Selected Items(s)** dialog box will be displayed. Next, select the required member in the main window; the selected member number will be displayed in the **Selected Items(s)** dialog box. Next, choose the **Done** button; the **Add New : Load Items** dialog box will be displayed again. You will notice that the selected member number will be displayed in the **Member** box. You can deselect any member by clearing the check box displayed corresponding to the member number.

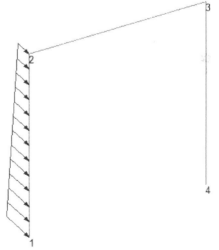

Figure 6-8 *Hydrostatic load applied on the structural member*

Now, in the **Force** area, specify the load values along with the minimum and maximum global axis in the **W1** and **W2** edit boxes, respectively. Specify the direction of loading by selecting the corresponding radio button in the **Direction** area. Next, choose the **Add** button to add the load and assign it to the desired structural members.

Prestress and Poststress

When pre stress load is applied on a structural member, its effect gets transmitted to all other connected members. In case of post stress load, the effect remains on the member itself and is not to be transmitted to any of the connecting members. To define pre stress/post stress load, select the **Pre/Post Stress** load option available under the **Member Load** option; the **Pre/Post Stress** page will be displayed in the right pane of the **Add New : Load Items** dialog box. In this page, define the loading type by selecting the **Prestress** or **Poststress** radio button in the **Type** area. In the **Load** area, specify the force value in the **Force** edit box. In the **Eccentricity Distances** area, specify the eccentricity distances with respect to the centre of gravity of the cross-section at the start, middle, and end of the member in the corresponding edit boxes. Next, choose the **Add** button to add the load and then assign it to the desired structural member.

Note
*You need to define the pre stress and post stress loads in two different load cases in the **Load & Definition** window.*

The command for assigning the prestress and poststress load is given next.

```
MEMBER PRESTRESS/POSTSTRESS LOAD
MEMBER-LIST FORCE {f₁} ES {d₁} EM {d₂} EE {d₃}
```

In the above command, the first line represents the load to be selected: prestress or post stress. In the next line, the **MEMBER -LIST** command represents the list of associated members. The **FORCE** command represents the load to be applied and $\{f_1\}$ represents the force value. **ES**, **EM**, and **EE** represent the eccentricity distances at the start, middle, and end of the member, respectively. Similarly, $\{d_1\}$, $\{d_2\}$, and $\{d_3\}$ represent the values of the eccentricity distances. Note that you do not need to include the parentheses while using the commands in the STAAD Editor window.

Fixed End Loads

Fixed end loads are applied at the member ends and are an alternative for the member loads. To define fixed end loads, select the **Fixed End** load option under the **Member Load** node in the left pane of the **Add New : Load Items** dialog box; the **Fixed End** load page will be displayed in the right pane. In this page, specify the force in x, y, and z directions and moments in x, y, and z direction in their corresponding edit boxes in the **Start Node** and **End Node** areas, respectively. Choose the **Add** button to add the load to the **Load & Definition** window and then assign it at the appropriate member ends.

Note
In this chapter, you need to download the c06_Staad_connect.zip file for the examples from http://www.cadcim.com. The path of the file is as follows: Textbook > Civil/GIS > STAAD.Pro > Exploring Bentley STAAD.Pro CONNECT Edition.

Example 1

In this example, you will open the *c06_staad_connect_ex1_start.std* file. Next, you will define some of the member loads and then assign them to the structural members.

Steps to complete this example are given below:

Step 1: Open the file *c06_staad_connect_ex1_start.std* in STAAD.Pro; the model is displayed in the main window.

Step 2: Invoke the **Loading** tab; the **Load & Definition** window is displayed in the right area of the interface.

Step 3: In this window, select the **Load Cases Details** as the title node and then choose the **Add** button; the **Add New : Load Cases** dialog box is displayed.

Step 4: In this dialog box, enter **Dead Load** in the **Title** edit box and retain the default settings. Next, choose the **Add** button and then choose the **Close** button. The load case is added in the **Load & Definition** window. Similarly, add another load case as **Live Load** and then close the **Add New : Load Cases** dialog box.

Step 5: Select the **Dead Load** case in the window and choose the **Add** button; the **Add New : Load Items** dialog box is displayed.

Step 6: In this dialog box, expand the **Member Load** node; the **Uniform Force** page is displayed. In this page, specify the value **-15** in the **W1** edit box and specify **1** and **2** in the **d1** and **d2** edit boxes, respectively and choose the **Add** button to add the load.

Step 7: Select the **Concentrated Force** in the left pane of the **Add New : Load Items** dialog box; the **Concentrated Force** page is displayed in the right pane.

Step 8: In this page, specify the value **-10** in the **P** edit box in the **Force** area and choose the **Add** button to add the load.

Step 9: Select the **Concentrated Moment** in the left pane of the dialog box; the **Concentrated Moment** page is displayed in the right pane. In this page, specify the value **22** in the **P** edit box and select the **GZ** radio button in the **Direction** area. Choose the **Add** button to add the moment and then close the dialog box.

Step 10: Now, select the load which was added first and then select the **Use Cursor To Assign** radio button. Next, choose the **Assign** button. Next, click on the members 2 and 3, refer to Figure 6-9; the load is assigned on these members.

Step 11: Repeat the procedure followed in step 10 to assign the second and third load on the 11 and 12 members, respectively.

Step 12: Repeat the previous steps to add the **Uniform Force** of **20kN/m** in the **Live Load** case and assign it to the members 2, 3, 6, 8, 11, and 12. Figure 6-10 shows the live load applied on the structure.

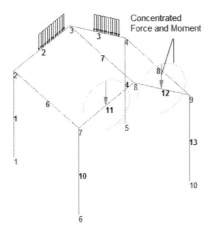

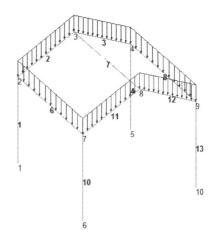

Figure 6-9 *Dead load applied on the structure*   ***Figure 6-10*** *Live load applied on the structure*

Step 13: Choose the **Save As** option from the **File** menu; the **Save As** dialog box is displayed. In this dialog box, specify the name *c06_staad_connect_ex1* in the **File name** edit box and save it at an appropriate location. Close the file by choosing the **Close** option from the **File** menu.

Area Load

Area load is applied on closed panel structures and is used for one way distribution. In STAAD.Pro, to define area load, you need to define pressure intensity and the direction in which load will be applied. The program will then automatically calculate the tributary area and the loads will be applied on the individual members of the closed panel as uniformly distributed loads. To define area load, select the **Area Load** option from the left pane of the **Add New : Load Items** dialog box; the **Area** page will be displayed. In this page, specify the pressure intensity in the **Pressure** edit box. Specify the direction by selecting the required radio button in the **Direction** area. By default, the direction will be in the local z direction. Next, choose the **Add** button to add the load and then assign it to the structure. After assigning the load to the closed panel structure, you will not be able to view the applied loads in the main window. You can view the applied area load on the structure only after the analysis has been performed. After the analysis, the loads will be applied on the members as member loads. You can also use the commands given next to define and assign area load.

```
AREA LOAD
MEMBER-LIST ALOAD {f} DIRECTION-SPECIFICATION
```

Floor Load

Floor load is used for two way distribution and is applied on closed panel structures. Figure 6-11 shows the floor load acting on a structure. To define floor load, define a load case as discussed earlier. Select the defined load case and choose the **Add** button; the **Add New : Load Items** dialog box will be displayed. In this dialog box, select the **Floor Load** option in the left pane; the **Floor** page will be displayed in the right pane. The parameters in this page are discussed next.

In the **Floor** page, the **YRANGE** radio button will be selected by default and the related options will be displayed in the **Range** area. In the **Range** area, define **Y Range**, **X Range**, and **Z Range** by specifying the appropriate values in their corresponding **Minimum** and **Maximum** edit boxes. Specify the pressure intensity in the **Pressure** edit box in the **Load** area. Next, specify the direction of loading by selecting the corresponding radio button in the **Direction** area. Select the **One Way Distribution** check box for one way distribution of loading. In this case, loads will be distributed on the longer side of the panel. You can also select the **Group** radio button in the **Floor** page. In this case, first you have to define the members forming the closed panel as the **Floor** group. To create a group, first select the members forming the closed panel in the main window. Next, choose the **Create New Group** option from the **Tools** menu; the **Define Group Name** dialog box will be displayed. Specify the desired name in the **Group Name** text box and select the **Floor** option from the **Select Typ**e drop-down list. Choose the **OK** button; the **Create Group** dialog box will be displayed. In this dialog box, choose the **Associate** button and then choose the **Close** button to close the dialog box. Now, select the **Group** radio button in the **Add New : Load Items** dialog box; the floor groups will be displayed in the **Member Group** list. Next, select the required group from this list box. Choose the **Add** button to add the loading in the **Load & Definition** window. The floor load will be automatically assigned to the structure.

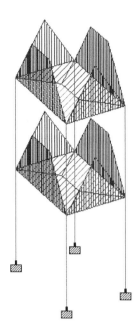

Figure 6-11 Floor load acting on the structure

Note
After assigning the loads to a structure, when you click on the defined loads in the Load & Definition window, the assigned load will be highlighted in blue color in the main window.

Example 2

In this example, you will open the *c06_staad_connect_ex2_start.std* file. Next, you will define floor load and assign it to the structure.

Steps to complete this example are given below:

Step 1: Open the file *c06_staad_connect_ex2_start.std* in STAAD.Pro; the model is displayed in the main window.

Step 2: Invoke the **Loading** tab; the **Load & Definition** window is displayed.

Step 3: Select the **Load:CASE 1** under the **Load Cases Details** head and choose the **Edit** button; the **Edit : Primary Load** dialog box is displayed. In the **Title** edit box, edit the name to **Floor Load** and then choose the **Change** button and close the dialog box.

Step 4: Select the created load case and choose the **Add** button; the **Add New : Load Items** dialog box is displayed.

Step 5: In this dialog box, the **Selfweight Load** page is displayed in the right pane.

Step 6: In the **Add New : Load Items** dialog box, expand the **Floor Load** node; the **Floor** page is displayed in the right pane. In this page, specify the values for the parameters, as shown in Figure 6-12. Choose the **Add** button to add the load.

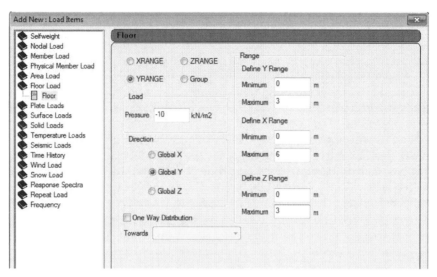

Figure 6-12 *Values specified in the **Add New : Load Items** dialog box*

Step 7: In the **Floor** page of the **Add New : Load Items** dialog box, specify the values for the second floor load, as shown in Figure 6-13. Choose the **Add** button to add the load and close the dialog box by choosing the **Close** button.

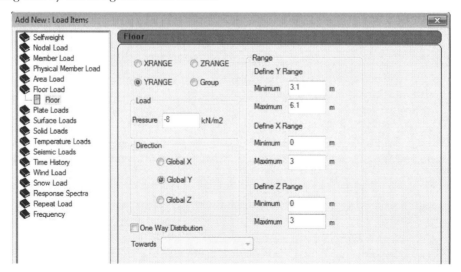

Figure 6-13 *Values specified in the **Add New : Load Items** dialog box for the 1st floor*

Step 8: Now, click on both the loads one by one in the **Load & Definition** window to view the applied loads on the structure. Figure 6-14 shows the applied floor loads on the structure.

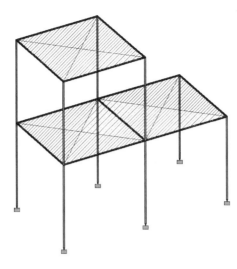

Figure 6-14 *Floor loads applied on the structure for the 2nd floor*

Step 9: Choose the **Save As** option from the **File** menu; the **Save As** page is displayed. In this dialog box, specify the name *c06_staad_connect_ex2* in the **File Name** edit box and save it at an appropriate location. Close the file by choosing the **Close** option from the **File** menu. The command for the assigned floor load is given next.

```
LOAD 1 LOADTYPE None TITLE Floor Load
SELFWEIGHT Y -1 LIST 1 TO 5 7 8 10 12 TO 24
FLOOR LOAD
YRANGE 0 3 FLOAD -10 XRANGE 0 6 ZRANGE 0 3 GY
YRANGE 3.1 6.1 FLOAD -8 XRANGE 0 3 ZRANGE 0 3 GY
```

Plate Loads

Plate loads can only be used in case of plates. To define plate loads, click on the **Plate Loads** node in the **Add New : Load Items** dialog box; the **Pressure On Full Plate** load option will be selected in the left pane and the **Pressure On Full Plate** page will be displayed in the right pane of the dialog box. Various types of plate load options in this dialog box are discussed next.

Pressure On Full Plate

The **Pressure On Full Plate** load option is used to define the load which will act on the full surface of the plate element. In the **Pressure On Full Plate** page of the **Add New : Load Items** dialog box, you need to specify the pressure intensity in the **W1** edit box in the **Load** area. In the **Direction** area, you can specify the local or global direction by selecting the required radio button. Next, choose the **Add** button to add the load in the **Load & Definition** window and then close the dialog box. You can assign the added load to the desired plate using any of the assigning options. By default, pressure will be applied in the local z direction. Figure 6-15 shows the pressure acting on a plate in the local z direction. The command to define and assign this load is given next.

```
ELEMENT LOAD
   ELEMENT-LIST PRESSURE DIRECTION-SPEC P₁
```

Concentrated Load

Concentrated load is used to define the load acting at a particular point in a plate. You can define this load by selecting the **Concentrated Load** option available under the **Plate Loads** node in the **Add New : Load Items** dialog box; the **Concentrated Load** page will be displayed. In this page, specify the force value in the **Force** edit box in the **Load** area. Specify the location of load on the plate by specifying the distance of coordinates from the origin in the **X** and **Y** edit boxes. Next, specify the direction of loading by selecting the required radio button in the **Direction** area. Add the defined load by choosing the **Add** button and then assign it to the desired plate. Figure 6-16 shows the concentrated load acting on a plate in the local z direction. The command for assigning this load is given next.

```
ELEMENT LOAD
ELEMENT-LIST PRESSURE DIRECTION-SPEC P₁ X Y
```

In the above command, P_1 represents the pressure value. **X** and **Y** represent the distance of coordinates of a point of load application from the origin.

Note
*To apply load on a node in a plate, you can use the **Node** option under the **Nodal Load** node as explained earlier in this chapter.*

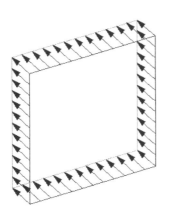
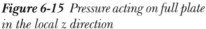
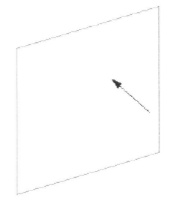

Figure 6-15 Pressure acting on full plate in the local z direction   *Figure 6-16* Concentrated load acting on a plate

Partial Plate Pressure Load

You can use this option to define the load acting partially on the plate element in the area defined by the user. To define this load, select the **Partial Plate Pressure Load** option from the left pane in the **Add New : Load Items** dialog box; the **Partial Plate Pressure Load** page will be displayed in the right pane. In this page, specify the pressure intensity in the **W1** edit box in the **Load** area. Specify the coordinates for the area in the **X1, Y1, X2,** and **Y2** edit boxes. Next, specify the direction of loading in the **Direction** area. Add and assign the load to the plate in the same way as discussed earlier. The command to define and assign this load is given next.

```
ELEMENT LOAD
ELEMENT-LIST PRESSURE DIRECTION-SPEC P₁ X₁ Y₁ X₂ Y₂
```

In the above commands, X_1, Y_1, X_2, and Y_2 represent the coordinate value for the enclosed area in the plate.

Trapezoidal Load

The trapezoidal load type is used to apply the trapezoidally varying pressure on a plate. To define this load, select the **Trapezoidal** option from the left pane of the **Add New : Load Items** dialog box; the **Trapezoidal** page will be displayed in the right pane of the dialog box. In this page, first you will specify the direction in which the pressure will act. To do so, select the required radio button in the **Direction of pressure** area. By default, the **Local Z** radio button is selected which shows that the pressure will be applied normally to the plate element in the local z direction. Next, you will specify the direction in which pressure will vary by selecting the **X** or **Y** radio button in the **Variation along element** area. If you select the **Joint** radio button from this area then you need to specify the pressure value for each node in the corresponding edit box. Specify the pressure intensity at the start and end in the **Start (f1)** and **End (f2)** edit boxes. Now, choose the **Add** button to add the load and then assign it to the desired plates. You can use the command given next to assign the trapezoidal load. Figures 6-17 and 6-18 show partial plate pressure and trapezoidal load acting on a plate.

```
ELEMENT LOAD
ELEMENT-LIST TRAP {GX, GY, OR GZ} {X, Y, OR JT} f₁ f₂
```

In the above command, the **TRAP** command is used in case of trapezoidal load. **GX**, **GY**, or **GZ** represent the direction in which pressure will be applied. **X**, **Y**, or **JT** represent the direction in which pressure will vary. f_1 and f_2 represent the pressure values.

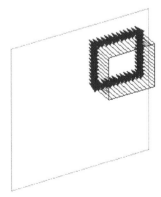 

Figure 6-17 *Partial plate pressure load acting on a plate*

Figure 6-18 *Trapezoidal load acting on a plate*

Hydrostatic Load

Using hydrostatic load, you can define hydrostatic pressure on plate elements. To define hydrostatic load, select the **Hydrostatic** load option from the left pane of the **Add New : Load Items** dialog box; the **Hydrostatic** page will be displayed in the right pane. In this page, you need to specify the values for the parameters in the same way as discussed earlier for the member loads. This load is converted to trapezoidal load on the plate elements. Figure 6-19 shows the hydrostatic load acting trapezoidally on a plate. The command for the hydrostatic load will be the same as for the trapezoidal load mentioned before.

Element Joint Load

You can define varying pressure on each joint in a plate element using the Element Joint Load. To do so, select the **Element Joint Load** option from the left pane of the **Add New : Load Items** dialog box; the **Element Joint Load** page will be displayed. In this page, first you have to specify the plate type whether it is three or four noded. To do so, select the **Three Noded Facet** or **Four Noded Facet** radio button in the **Joint Load Data** area. Next, specify the node numbers and the corresponding pressure intensity in their respective columns in the table. Specify the direction of pressure acting on the element in the **Direction** area. Now, choose the **Add** button; the load will be added to the **Load & Definition** window. The Element Joint Load will be assigned automatically to the specified plate. Figure 6-20 shows the element joint load acting on a plate element. The command for assigning this load is given next.

```
ELEMENT LOAD JOINT
n₁ n₂ n₃ n₄ FACETS f PRESSURE p₁ p₂ p₃ p₄
```

In the above command, n_1, n_2, n_3, and n_4 represent the four nodes of the plate on which load will be applied and p_1, p_2, p_3, and p_4 represent the pressure values at the four nodes.

Figure 6-19 Hydrostatic load acting on a plate

Figure 6-20 Element joint load acting on a plate

Example 3

In this example, you will open the *c06_staad_connect_ex3_start.std* file. Next, you will define pressure acting on a full plate and then assign it to the plate.

Steps required to complete this example are given below:

Step 1: Open the file *c06_staad_connect_ex3_start.std* in STAAD.Pro; the model is displayed in the main window, as shown in Figure 6-21.

Step 2: Next, click in the main window and press SHIFT+P to view the plate numbers.

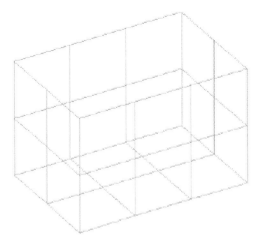

Figure 6-21 *Model displayed in the Main Window*

Step 3: Invoke the **Loading** tab; the **Load & Definition** window will be displayed.

Step 4: In this window, select the **Load Cases Details** and choose the **Add** button; the **Add New : Load Cases** dialog box appears. Create a load case with the **Loading Type** as **None** and **Title** as **Plate Load** and then choose the **Add** button and **Close** the dialog box. Select the created load case and choose the **Add** button; the **Add New : Load Items** dialog box is displayed.

Step 5: In this dialog box, expand the **Plate Loads**; the **Pressure on Full Plate** load is highlighted in the left pane of the window and the **Pressure on Full Plate** page is displayed in the right pane.

Step 6: In this page, specify the value **1** in the **W1** edit box in the **Load** area and choose the **Add** button to add the load and then close the dialog box.

Step 7: Select the added load and choose the **Assign** button in the **Assignment Method** area. Click in the main window and using the left and right arrow buttons, rotate the model.

Step 8: Next, click on the plate 1; the pressure load is assigned to plate 1, refer to Figure 6-22.

Step 9: Similarly, assign the pressure load to all the plates in the first row, refer to Figure 6-23.

Step 10: Choose the **Save As** option from the **File** menu; the **Save As** dialog box is displayed. In this dialog box, specify the name *c06_staad_connect_ex3* in the **File name** edit box and save it at an appropriate location. Close the file using the **Close** option in the **File** menu.

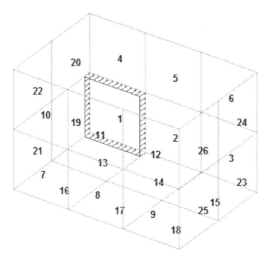

Figure 6-22 Pressure load assigned to the plate 1

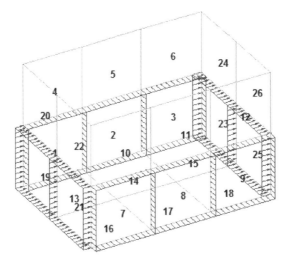

Figure 6-23 Pressure load applied on all the first row plates

Surface Loads

Surface loads allows you to define uniform pressure on surface elements in various ways. This will be discussed later in this chapter. To define a surface load, expand the **Surface Loads** node in the left pane of the **Add New : Load Items** dialog box; the **Concentrated Load** option will be selected in the left pane and the **Concentrated Load** page will be displayed in the right pane of the dialog box. The options in this page are discussed next.

Concentrated Load

Concentrated load will be applied at a particular point on the surface element. You need to specify the x and y coordinates of the point where the load will be applied. Note that the coordinates of this point will be taken with respect to local coordinate system and the first node of the surface element will be taken as the origin. In the **Load** area of the **Concentrated Load** page, specify the force value in the **Force** edit box. Next, specify the distance of the point from the first node in the **X** and **Y** edit box. Next, specify the direction of load by selecting the required radio button in the **Direction** area. After specifying the parameters, choose the **Add** button; the load will be added to the **Load & Definition** window. Now, you can assign the added load to the desired surface element. Figure 6-24 shows the concentrated load applied on a surface element.

Pressure on Full Surface Load

The pressure on full surface can be defined using the **Pressure on Full Surface** load option. To define this load, click on the **Pressure on Full Surface** load option available under the **Surface Loads** node in the left pane of the **Add New : Load Items** dialog box; the **Pressure on Full Surface** page will be displayed in the right pane of the dialog box. In the **Load** area of this page, you will specify the pressure intensity in the **W1** edit box. Next, specify the direction in which load will act by selecting the required radio buttons such as **Local Z**, **GX**, **GY**, or **GZ** in the **Direction** area. Next, choose the **Add** button to add the load to the **Load & Definition** window. Now, you can assign the added load to the desired surface element. Figure 6-25 shows the pressure applied on the full surface element.

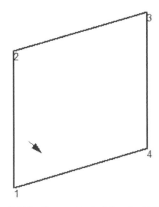

Figure 6-24 *Concentrated load applied on surface*

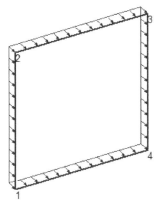

Figure 6-25 *Uniform pressure applied on surface*

Partial Surface Pressure Load

Pressure can also be applied on a surface element partially but in a uniform manner. To do so, click on the **Partial Surface Pressure Load** option in the left pane of the **Add New : Load Items** dialog box; the **Partial Surface Pressure Load** page will be displayed in the right pane of the dialog box. In the **Load** area of this page, specify the pressure intensity in the **W1** edit box. Next, specify the coordinates for the enclosed surface on which load will be applied in the **X1**, **X2**, **Y1**, and **Y2** edit boxes. Specify the direction of loading in the **Direction** area and then choose the **Add** button to add the load to the **Load & Definition** window. Assign the added load to the surface element. Figure 6-26 shows the partial surface pressure load applied on the surface.

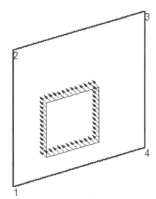

Figure 6-26 *Partial surface pressure load applied on a surface*

Partial Surface Trapezoidal Load

When you need to apply the pressure loading on a surface partially in a non-uniform manner, you can use the **Partial Surface Trapezoidal Load** option. To do so, click on the **Partial Surface Trapezoidal Load** option in the left pane of the **Add New : Load Items** dialog box; the **Partial Surface Trapezoidal Load** page will be displayed in the right pane of the dialog box. In the **Direction of pressure** area, specify the direction of loading by selecting the required radio buttons. In the **Load Position** area, specify the coordinates to define the load area. Next, specify the pressure values at the four corners of the loaded area in the edit boxes in the **Variation along element** area. Next, choose the **Add** button to add the load and then assign it to the surface. Figure 6-27 shows the partial surface trapezoidal load applied on a surface.

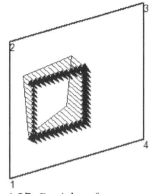

Figure 6-27 *Partial surface trapezoidal load applied on the surface element*

 Note
*The commands used for all the surface loads will be the same as for the plate loads. For surface loads, instead of using the **PLATE LOAD** command, you need to start with the **SURFACE LOAD** command in the **STAAD Editor** window.*

Solid Loads

Solid load is used to define uniform or varying pressure on the faces of a solid element. To define solid loads, click on the **Solid Loads** node in the left pane of the **Add New : Load Items** dialog box; the **Solid pressure load** option is selected and the corresponding page will be displayed in the right pane of the dialog box. In this page, select the required face number on which the load will be applied from the **Face number** drop-down list. Next, specify the pressure acting at each node in the **Node 1**, **Node 2**, **Node 3**, and **Node 4** edit boxes. Select the required radio button in the **Direction** area to specify the direction of pressure. Next, choose the **Add** button to add the load and then assign it to the required solid element. Figure 6-28 shows the load applied on a solid element.

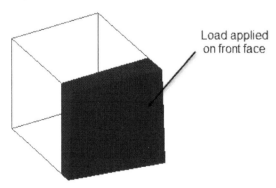

Load applied on front face

Figure 6-28 *Pressure applied on the front face of a solid element*

Temperature Loads

In a structural member, temperature differs throughout the member length which causes thermal stress/strain in a building. In STAAD.Pro, to define temperature load, click on the **Temperature Loads** node in the left pane of the **Add New : Load Items** dialog box; the **Temperature** load option will be selected in the left pane and the **Temperature** page will be displayed in the right pane of the dialog box. In this page, specify the value for the axial elongation in the **Temperature Change for Axial Elongation** edit box. Similarly, specify the required values in the **Temperature Differential from Top to Bottom** and **Temperature Differential From Side to Side (Local Z)** edit boxes. Next, choose the **Add** button to add the load and then assign it to the required member or element.

In the left pane of the **Add New : Load Items** dialog box, the **Strain** load option is also available under the **Temperature** load node. To specify a strain load, select the **Strain** option available under the **Temperature Loads** node in the left pane of the dialog box; the **Strain** page will be displayed in the right pane. In this page, specify the value for initial elongation or shrinkage caused in the member in the **Initial Axial Elongation (+) or Shrinkage (-)** edit box. Choose the **Add** button to add the load and then assign it to the structure.

Seismic Loads

When an earthquake occurs, a building is subjected to inertia forces acting in opposite direction of earthquake acceleration, which further subjects the building to dynamic motion. This inertia force is known as seismic load.

In STAAD.Pro, to define seismic load, click on the **Seismic Loads** node in the left pane of the **Add New : Load Items** dialog box; the **Factor & Direction** option will be selected in the left pane and the **Factor _Direction** page will be displayed in an inactive state in the right pane of the dialog box and it will look similar to the one shown in Figure 6-29. The options in this page are disabled because no seismic load has been defined yet. The method of defining seismic load is discussed next.

Defining Seismic Load

In STAAD.Pro, seismic load is defined to perform the dynamic analysis using various codes. To define seismic load, in the **Load & Definition** window, expand the **Definitions** node. Next, select the **Seismic Definitions** option and choose the **Add** button; the **Add New : Seismic Definitions** dialog box will be displayed, as shown in Figure 6-30.

In this dialog box, select the required code from the **Type** drop-down list. Select the **Include Accidental Load** check box to calculate the accidental torsion component as per the selected code. The various parameters associated with the selected code will be displayed in the table of the **Seismic Parameters** page. For example, if you have selected the **UBC 1997** code, then the related parameters will be displayed in a table, as shown in Figure 6-30. In this table, you need to specify the values for the given parameters.

After specifying the values for the parameters, choose the **Add** button; the seismic parameters will be added and the **Self Weight** page will be displayed in the **Add New : Seismic Definitions** dialog box, refer to Figure 6-31. Now, you will specify the structural weight for calculating the Base Shear. The options used for defining the structural weight are discussed next.

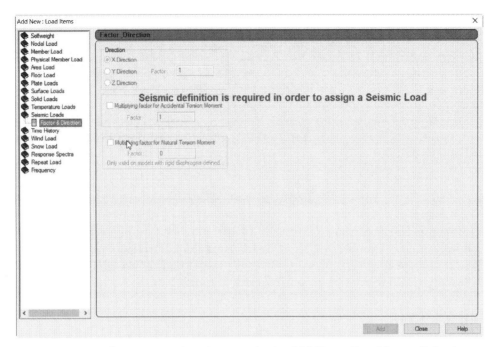

Figure 6-29 The **Factor _Direction** *page in the* **Add New : Load Items** *dialog box*

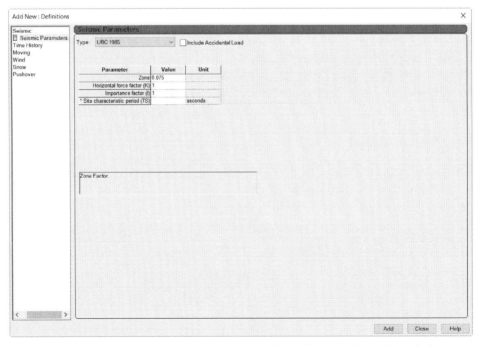

Figure 6-30 The **Seismic Parameters** *page in the* **Add New : Seismic Definitions** *dialog box*

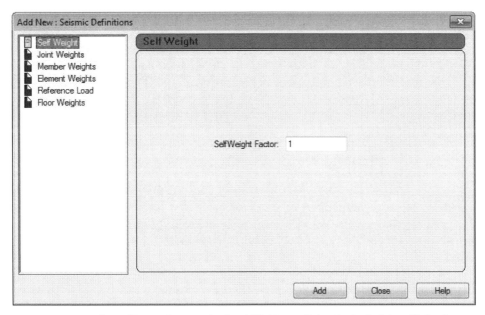

Figure 6-31 The Self Weight page in the Add New : Seismic Definition dialog box

Self Weight

In the **Self Weight** page displayed, specify the self weight factor in the **SelfWeight Factor** edit box and choose the **Add** button; the self weight load will be added under the **Seismic Definition** node in the **Load & Definition** window.

Joint Weights

Select the **Joint Weights** node in the left pane of the dialog box; the **Joint Weights** page will be displayed in the right pane of the dialog box. In the **Joint Weights** page, specify the concentrated load acting on the joint in the **Joint Weight** edit box. Next, choose the **Add** button; the load will be added under the **Seismic Definition** node in the **Load & Definition** window.

Member Weights

Next, select the **Member Weights** node in the left pane of the dialog box; the **Member Weights** page will be displayed in the right pane. In this page, you will specify the distributed and concentrated weights acting on the member. Select the loading type such as **CON** for concentrated load or **UNI** for uniform load from the **Loading Type** drop-down list. The loading parameters will be displayed according to the selected loading type. Specify the load intensity and the location of the load and then choose the **Add** button to add the load in the **Load & Definition** window.

Element Weights

In case of floor slabs and other structural models which consists of plate elements, you need to define the element weight. To do so, select the **Element Weights** node in the left pane of the dialog box; the **Element Weights** page will be displayed in the right pane of the dialog box. Specify the pressure intensity in the **Pressure** edit box and choose the **Add** button; the load will be added under the **Seismic Definition** node in the **Load & Definition** window.

Reference Load

In seismic loading, instead of individually defining self weight, member weights, joint weights, and element weights for structural weight, you can define the structural weight by adding the reference load. To do so, select the **Reference Load** node in the left pane of the dialog box; the **Reference Load** page will be displayed in the right pane of the dialog box, as shown in Figure 6-32. Now, you can see from the figure that a primary load case is required to add a reference load. To do so, close this dialog box first and select the **Reference Load Definitions** node in the **Load & Definition** window and choose the **Add** button; the **Add New : Reference Load Definitions** dialog box will be displayed. Specify the title, number, and loading type in this dialog box and after choosing the **Add** button, close the dialog box. Next, select the added reference load and choose the **Add** button; the **Add New : Reference Load Items** dialog box will be displayed. In this dialog box, you will define and add the loads which are similar to primary loads. Next, assign the added load to the structure. After assigning the loads, invoke the **Add New : Seismic Definitions** dialog box and then select **Reference Load** in the left pane of the dialog box; the **Reference Load** page will be displayed, as shown in Figure 6-33. Now, in this dialog box, select the added reference load case from the **Available Load Cases** list box and choose the **Forward** button to move it to the **Referenced Load** list box. Next, specify the required factor to be multiplied with the reference load in the **Factor** column. Select the direction from the **Along** drop-down list and choose the **Add** button to add the reference load. Next, close the dialog box.

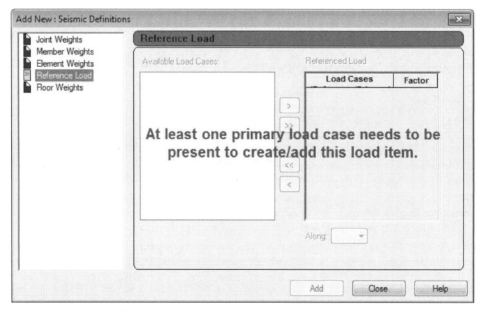

*Figure 6-32 The **Reference Load** page in the **Add New : Seismic Definitions** dialog box*

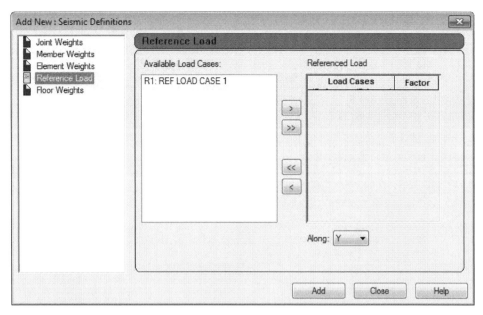

Figure 6-33 The **Reference Load** *page after adding a primary load case*

Floor Weights

You can define floor weights in case of a structure consisting of beams and columns instead of floor slab. To define floor weight, select the **Floor Weights** load option in the left pane of the dialog box; the **Floor Weights** page will be displayed in the right pane of the dialog box. Specify the value for load and the load location, as discussed in the **Floor Load** section. Choose the **Add** button to add the load in the **Load & Definition** window.

After defining seismic loads, you need to define the factor and global direction in which the load will be applied. To do so, invoke the **Add New : Load Items** dialog box and then select the **Seismic Loads** in the left pane of the dialog box; the **Factor_Direction** page will be displayed in the right pane of the dialog box, refer to Figure 6-34. Specify the direction in which load will be applied by selecting the **X Direction**, **Y Direction**, or **Z Direction** radio button in the **Direction** area. Specify the multiplying factor in the **Factor** edit box. If accidental torsion is included then select the **Multiplying factor for Accidental Torsion Moment** check box and specify the value in the **Factor** edit box. If natural torsion moment is included then select the **Multiplying factor for Natural Torsion Moment** check box and specify the value in the **Factor** edit box. Next, choose the **Add** button to add the load and close the dialog box.

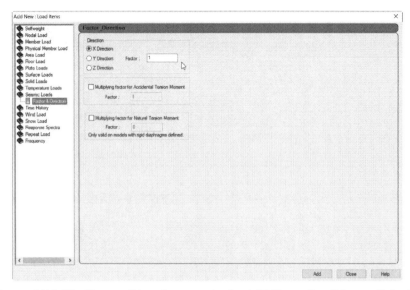

*Figure 6-34 The **Factor_Direction** page in the **Add New : Load Items** dialog box*

Example 4

In this example, you will open the *c06_staad_connect_ex2.std* file. Next, you will define seismic loading as per IBC 2006 for a space frame structure.

Steps required to complete this example are given below;

Step 1: Open the file *c06_staad_connect_ex2.std* in STAAD.Pro; the model is displayed in the main window, as shown in Figure 6-35.

Note

*If the **Assign Projects** dialog box is displayed, choose the **Cancel** button; the dialog box is closed and the file is loaded.*

Step 2: Choose the **Loading** tab; the **Load & Definition** window is displayed.

Step 3: Expand the **Definitions** node in the **Load & Definition** window and select the **Seismic Definitions** option. Next, choose the **Add** button; the **Add New : Seismic Definitions** dialog box is displayed.

Step 4: In the right pane of the dialog box, select the **IBC 2006/2009 ASCE 7-05** code from the **Type** drop-down list.

Step 5: Select the **Include Accidental Load** check box to calculate the accidental torsion component as per the IBC 2006 code.

Step 6: Specify **46201** in the **Value** column of the **Zip Code**, **10** in the **Value** column of **TL**, **1.2** in the **Value** column of **F**$_a$, and **1.7** in the **Value** column of **F**$_v$, refer to Figure 6-36. After specifying the parameters, choose the **Add** button to add the load definition.

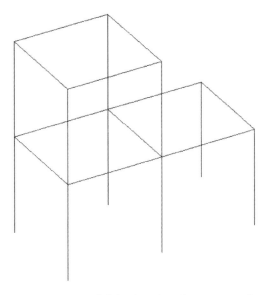

Figure 6-35 Model displayed in the main window

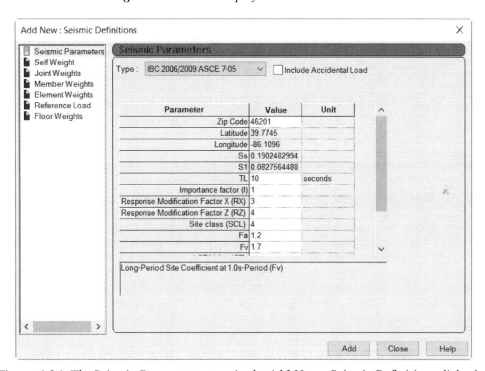

Figure 6-36 The *Seismic Parameters* page in the *Add New : Seismic Definitions* dialog box

Step 7: After adding the load definition, the **Self Weight** page is displayed in the left pane of the dialog box. Choose the **Add** button to add the self weight.

Step 8: Next, select the **Floor Weights** option in the left pane of the dialog box; the **Floor Weights** page is displayed in the right pane.

Step 9: Specify the pressure intensity and X, Y, and Z ranges, as given in Figures 6-37 and 6-38. Add both the floor weights one by one and then close the dialog box.

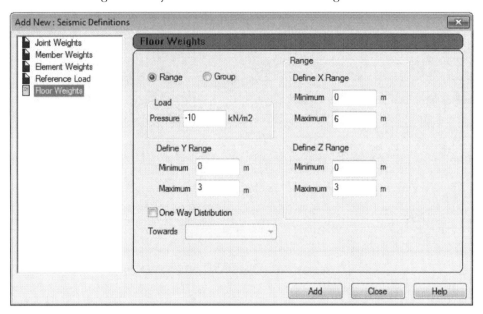

Figure 6-37 *Values for the floor weight 1*

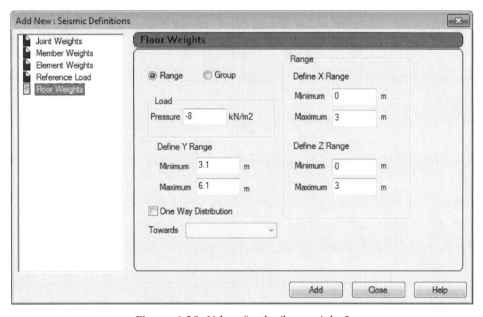

Figure 6-38 *Values for the floor weight 2*

Step 10: Next, expand the **Load Cases Details** node in the **Load & Definition** window, select the **FLOOR LOAD** sub node, and choose the **Add** button; the **Add New : Load Items** dialog box is displayed.

Step 11: In this dialog box, select **Seismic Loads** in the left pane of the dialog box; the **Factor_Direction** page is displayed in the right pane.

Step 12: Ensure that the **X Direction** radio button is selected and specify **1** in the **Factor** edit box in the **Direction** area. Choose the **Add** button to add the loading and close the dialog box. The **IBC LOAD X 1** is added under the **FLOOR LOAD** sub node in the **Load & Definition** window.

Step 13: Now, choose the **Editor** tool from the **Utilities** panel of the **Utilities** tab; the warning message box is displayed. Choose the **Save** button; the **STAAD Editor** window is displayed. In this window, select the **UBC LOAD X 1** command. Cut and paste this command below the **LOAD 1 LOADTYPE None TITLE FLOOR LOAD** command. The commands should be in the sequence given next. Choose the **Save** button and close the **STAAD Editor** window.

Step 14: Choose the **Save As** option from the **File** menu; the **Save As** page is displayed. In this dialog box, specify the name *c06_staad_connect_ex4* in the **File name** edit box and save it at an appropriate location. Close the file by choosing the **Close** option from the **File** menu.

Time History

Time history load provides for linear or non-linear analysis of dynamic structural response under loading which may vary according to the time. To use this type of load, choose the **Loading** tab; the **Load & Definition** window will be displayed. Select the **Load Cases Details** node in the list box. Next, choose the **Add** button; the **Add New: Load Cases** dialog box will be displayed. In the **Primary** page, specify the load case number in the **Number** edit box. Select load type from the **Loading Type** drop down list. Add title for the load case in the **Title** text box for reference. Select the **Title** text box for reference. Choose the **Add** button to add the load case and close the dialog box. Select the load case you add to the list box of the **Load & Definition** window, and choose the **Add** button; the **Add New : Load Items** dialog box will be displayed. Select the **Time History** node from the left pane of the **Add New : Load Items** dialog box; the **Time History** page will be displayed in the right pane, as shown in Figure 6-39. The options in this page are inactive. To activate the options, you need to define the time history loading. This option is discussed next.

Defining Time History Load

To define time history load in the **Load & Definition** window, expand the **Definitions** node. Next, select the **Time History Definitions** sub-node and choose the **Add** button; the **Add New : Time History Definitions** dialog box will be displayed, as shown in Figure 6-40. The options in this dialog box are discussed next.

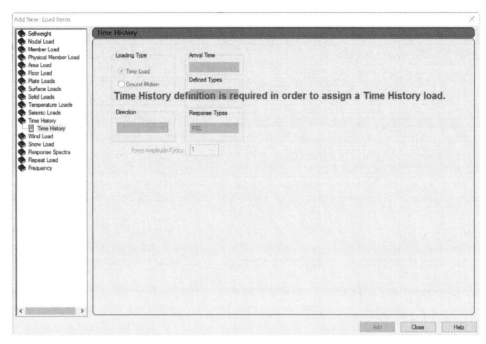

Figure 6-39 The **Time History** *page in the* **Add New : Load Items** *dialog box*

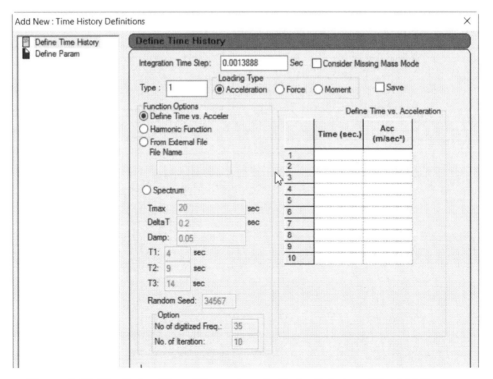

Figure 6-40 The **Define Time History** *page in the partial view of the* **Add New :**
Time History Definitions *dialog box*

Type

In the **Type** edit box, you will specify the number of the defined load type.

Loading Type

In the **Loading Type** area, you will define the type of function being used for defining time history loading by selecting the **Acceleration**, **Force**, or **Moment** radio button. On selecting any of the radio buttons, the selected options will be displayed in the table below. For example, if you have selected the **Force** radio button then the **Define Time vs Force** table will be displayed. Select the **Save** check box to create an external input file with the *.tmh* extension. This file will contain the history of displacements of each node of the structure.

Function Options

In the **Function Options** area, there are three types of functions. The options displayed in the **Loading Type** area depend on the type of function selected in this area. These functions are discussed next.

Define Time vs Acceleration: This radio button is selected by default in the **Function Options** area. In this function type, you need to specify the values for the time and corresponding acceleration, force, or moment. The time history forcing function will be plotted as per the specified values.

Harmonic Function: Harmonic function is used to define time history harmonic function. When you select this radio button, the related options will be displayed in the **Loading Type** area, as shown in Figure 6-41. In the **Others** area, you will define the sine or cosine function by selecting the **SINE** or **COSINE** radio button. Specify the frequency in cycles per second in the edit box below the **Frequency** radio button. You can also specify the revolutions per minute in the edit box below the **RPM** radio button. Similarly, specify the values for the **Amplitude**, **Phase**, **Cycles**, **Step**, and **SubDiv** in their corresponding edit boxes.

From External File: Select the **From External File** radio button to load an external file containing the time history data. On selecting this radio button, the **File Name** edit box will be activated. In this edit box, specify the file name which should not be more than 8 characters. The file should be placed in the same directory.

After defining the required time history loading function, choose the **Add** button; the definition will be added under the **Time History Definitions** node in the **Load & Definitions** window. Next, you need to define the parameters such as time step, damping, and arrival time for time history load, which is discussed next.

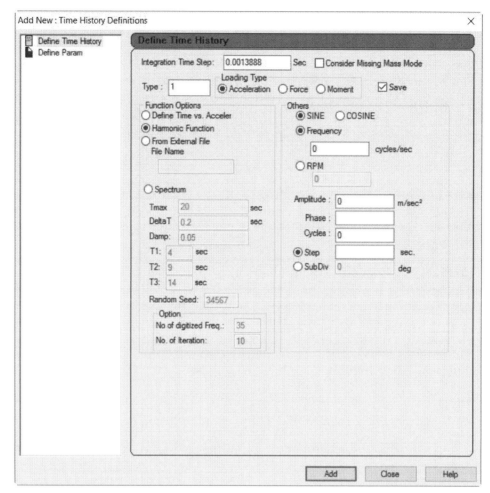

Figure 6-41 *Options displayed on selecting the **Harmonic Function** radio button*

Defining Parameters

In the left pane of the **Add New : Time History Definitions** dialog box, select the **Define Param** option; the **Define Param** page will be displayed in the right pane of the dialog box, as shown in Figure 6-42. In this page, you will define the time step, damping, and arrival time for time history load, which is discussed next.

Time Step

To specify the value of Integration time step of uncoupled equations, first select the **Time Step** check box and then specify a value in the edit box next to it.

Damping

In the **Damping** area, you will specify the damping ratio which will be applied to all the modes. To specify a single modal damping ratio, select the **Damping** radio button, if it is not selected by default. Next, specify a value in the edit box below the **Damping** radio button. The default value in this edit box is **0.05**.

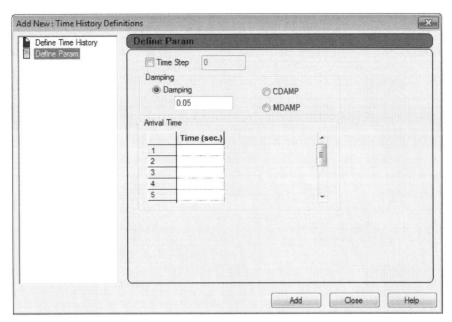

*Figure 6-42 The **Define Param** page in the **Add New : Time History Definitions** dialog box*

If the damping ratio is already defined in the type of material used in the structure, then you can use that value for the time history analysis. In such a case, you need to select the **CDAMP** radio button.

You can also use the individual damping ratios for individual modes. To do so, first you need to define the individual damping ratios. Individual damping ratios can be defined in the **Modal Damping** dialog box. This dialog box can be invoked by choosing the **Define Damping for Dynamics** option from the **Commands** menu. After defining the values, you need to select the **MDAMP** radio button in the **Define Param** page of the **Add New : Time History Definitions** dialog box.

Arrival Time
In this area, you will specify the possible arrival time of various dynamic load types in the **Time (sec.)** column. The arrival time is the time at which the load type will begin to act at the joint or at the base of the structure.

After defining the required time history loading function, choose the **Add** button; the defined parameters will be added under the **Time History Definitions** node in the **Load & Definitions** window.

Note
The arrival time and the time-force values for the load types are used to create the load vector required at each time step of the analysis.

After defining the time history load, you need to define the loading as primary load type. To do so, select the required load case, and choose the **Add** button; the **Add New : Load Items** dialog box will be displayed. In the left pane of the dialog box, select the **Time History** node; the **Time History** page will be displayed in the right pane, refer to Figure 6-43. The options in this page are discussed next.

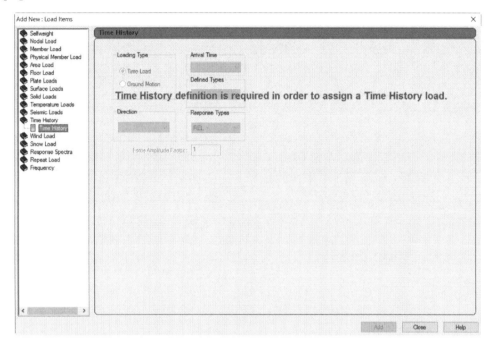

*Figure 6-43 The **Time History** page in the **Add New : Load Items** dialog box*

In the **Loading Type** area, select the **Time Load** radio button to apply the time history load to the joints in the structure. Select the **Ground Motion** radio button to apply the load at the base of the structure. From the drop-down list in the **Arrival Time** area, you need to select the arrival time at which the load begins to act. Select the required time from the drop-down list. In the **Direction** area, specify the global direction in which the load will be applied by selecting an appropriate option from the drop-down list. In the **Defined Types** area, select the previously defined type number from the drop-down list. Specify a value in the **Force Amplitude Factor** edit box. This value will be multiplied with the previously defined force or acceleration values.

Wind Load

Wind load allows you to define the parameters for automatic generation of wind load as joint load on the structure. To define this load, select the **Wind Load** node from the left pane of the **Add New : Load Items** dialog box; the **Wind Load** page will be displayed in the right pane, as shown in Figure 6-44. The options in this page are inactive. To activate the options, you need to define the wind load. The process to do so is discussed next.

Defining Wind Load

To define wind load, expand the **Definition** node in the **Load & Definition** window and then select the **Wind Definitions** option. Next, choose the **Add** button; the **Add New : Wind**

Definitions dialog box will be displayed. In this dialog box, the wind load type number and the comments will be specified by default in the **Type No** and **Comments** edit boxes, respectively. Choose the **Add** button to add the type and close the dialog box; **TYPE 1 : WIND 1** will be added under the **Wind Definitions** option in the **Load & Definition** window. Next, select the **TYPE 1 WIND 1** definition and choose the **Add** button; the **Intensity** page in the **Add New : Wind Definitions** dialog box will be displayed, as shown in Figure 6-45.

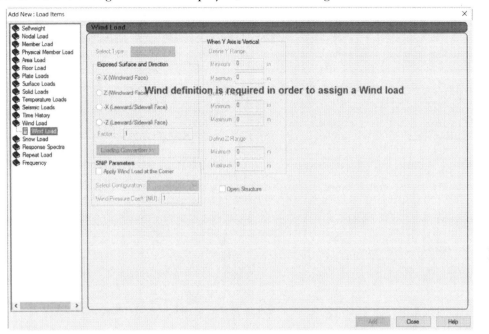

Figure 6-44 The **Wind Load** page in the **Add New : Load Items** dialog box

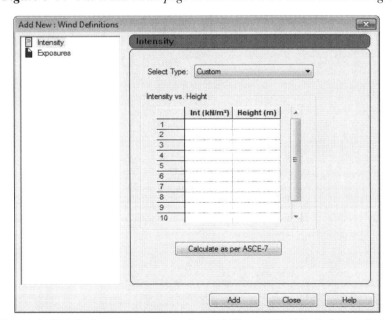

Figure 6-45 The **Intensity** page in the **Add New : Wind Definitions** dialog box

In this page, the **Intensity vs. Height** table is displayed where you need to enter the values for the wind intensity and the height above the ground. You can also calculate these values as per ASCE - 7. To do so, choose the **Calculate as per ASCE - 7** button; the **Common** page in the **ASCE - 7 : Wind Load** dialog box will be displayed, as shown in Figure 6-46. There are three pages displayed on choosing the options available in the left pane of the dialog box. The options in these pages are used to define the wind intensity. These pages are discussed next.

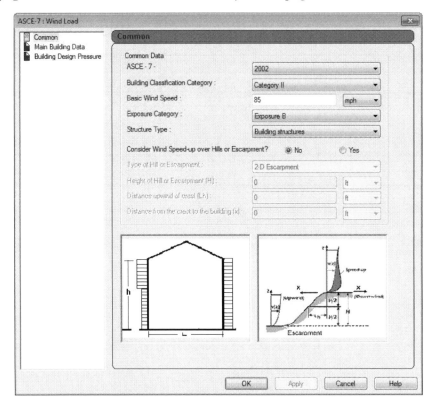

Figure 6-46 The ASCE - 7 : Wind Load dialog box

Common

In the **Common** page, you need to define the common data such as the code to be used, building category, wind speed, structure type, and so on. Select the required code type from the **ASCE - 7** drop-down list. Select the building category from the **Building Classification Category** drop-down list. Specify the wind speed value in the **Basic Wind Speed** edit box. Select the wind exposure category from the **Exposure Category** drop-down list. Specify the type of structure such as chimney, tanks, tower, and so on by selecting the appropriate option from the **Structure Type** drop-down list. If there are changes in the topography such as hills or escarpments, then select the **Yes** radio button for considering the wind speed over hills or escarpments. The options under the radio button become active. Next, select the type of the hill or the escarpment from the **Type of Hill or Escarpment** drop-down list. Specify the height of hill or escarpment in the **Height of Hill or Escarpment (H)** edit box. Specify the distance upwind of crest in the **Distance upwind of crest (Lh)** edit box.

Similarly, specify the distance from the crest to the building site in the **Distance from the crest to the building (x)** edit box. After specifying all the parameters, choose the **OK** button; the intensity and height values will be displayed in the **Intensity** page of the **Add New : Wind Definitions** dialog box.

Main Building Data

The intensity vs height values can also be defined using the options available in the **Main Building Data** page. To do so, choose the **Main Building Data** node in the left pane of the **ASCE - 7 : Wind Load** dialog box; the **Main Building Data** page will be displayed. In this page, you will specify the building height, building length along the direction of the wind, building length normal to the direction of wind, building natural frequency, damping ratio, and type of enclosure. These parameters will calculate the wind intensity according to the height. After specifying all the parameters, choose the **OK** button; the intensity and height values will be displayed in the **Intensity** page of the **Add New : Wind Definitions** dialog box.

Building Design Pressure

The options in this page are used to calculate the intensity vs height values. To do so, choose the **Building Design Pressure** node in the left pane of the **ASCE - 7 : Wind Load** dialog box; the **Building Design Pressure** page will be displayed in the right pane. To calculate the design wind pressure for the windward, leeward, or sidewall, select the corresponding radio button from the **Generate Wind Load On Wall** area. To include the gust effect factor, external pressure coefficient, and internal pressure coefficient, select the G, C_p, and GC_{pi} check boxes and then specify the value in the corresponding edit boxes. Next, choose the **OK** button; the wind intensity at various heights will be displayed in the **Intensity** page of the **Add New : Wind Definitions** dialog box.

Next, choose the **Add** button to add the load definition in the **Load & Definition** window. On doing so, the **Exposures** page will be displayed in the **Add New : Wind Definitions** dialog box. In this page, you have to specify the exposure factor value in the **Factor** edit box. This exposure value is the influence area of the wind load associated with the particular joints in the structure. Choose the **Add** button to add the exposure factor and then close the dialog box.

After defining the wind load, you need to define the loading as primary load type. To do so, select the required load case and choose the **Add** button; the **Add New : Load Items** dialog box will be displayed. In the left pane of the dialog box, select the **Wind Load** node; the **Wind Load** page will be displayed in the right pane, refer to Figure 6-47. The options in this page are discussed next.

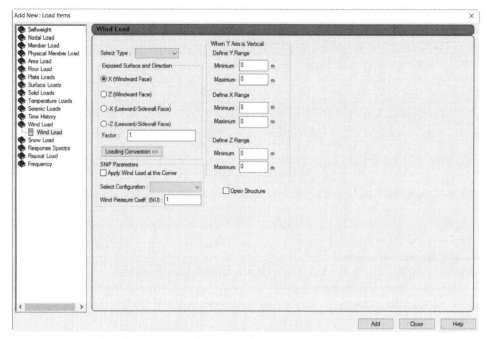

Figure 6-47 *The **Wind Load** page in the **Add New : Load Items** dialog box*

Select the previously defined wind load type in the **Select Type** drop-down list. In the **Direction** area, to specify the global direction in which the wind load is to be applied, select the **X, Z, -X,** or **-Z** radio button. Specify the multiplying factor to calculate the wind loads in the **Factor** edit box. In the **When Y Axis is Vertical** area, you will define the X, Y, and Z range of the structure. To generate wind load on open structures such as transmission towers, bridges, or any other open structure, select the **Open Structure** check box. By default, the structures are considered as closed structure. Next, choose the **Add** button; the load will be added under the selected load case and then close the **Add New : Load Items** dialog box. Now, click on the added load; the wind load will be displayed at the joints in the structure, refer to Figure 6-48.

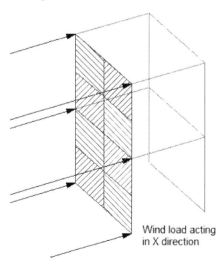

Wind load acting
in X direction

Figure 6-48 *Wind load acting in the X direction*

Example 5

In this example, you will open the *c06_staad_connect_ex5_start.std* file. Next, you will define wind loads for a space frame structure as per ASCE 7-2002.

The following steps are required to complete this example:

Step 1: Open the file *c06_staad_connect_ex5_start.std* in STAAD.Pro. Press Shift+N; the model is displayed in the main window, as shown in Figure 6-49.

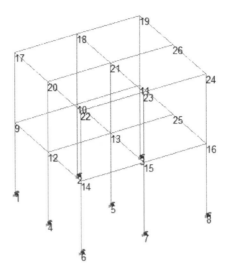

Figure 6-49 Model displayed in the main window

Note
*If the **Assign Projects** dialog box is displayed, choose the **Cancel** button; the dialog box is closed and the file is loaded.*

Step 2: Choose the **Loading** tab; the **Load & Definition** window is displayed. In the **Load & Definition** window, expand the **Definitions** node and select the **Wind Definitions** sub node.

Step 3: Next, choose the **Add** button; the **Add New : Wind Definitions** dialog box is displayed. In this dialog box, choose the **Add** button and then close the dialog box; the **TYPE 1 : WIND 1** node is added under the **Wind Definitions** sub node in the **Load & Definition** window.

Step 4: Select the **TYPE 1 : WIND 1** node and choose the **Add** button; the **Add New : Wind Definitions** dialog box is displayed. In this dialog box, choose the **Calculate as per ASCE-7** button; the **ASCE 7 : Wind Load** dialog box is displayed.

Step 5: In this dialog box, select the **Main Building Data** option in the left pane of the dialog box; the **Main Building Data** page is displayed in the right pane of the dialog box.

Step 6: Specify the parameters for automatic generation of the wind load, as shown in Figure 6-50.

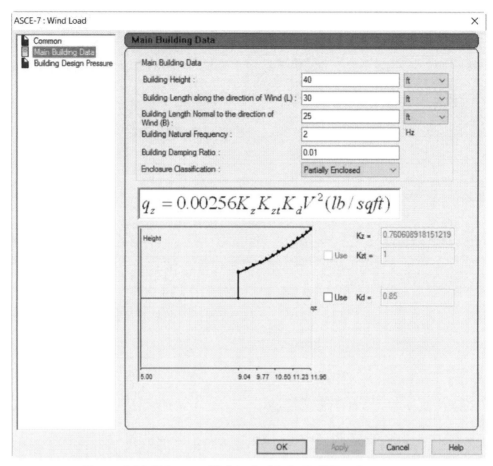

*Figure 6-50 Values specified in the **Main Building Data** page*

Step 7: Choose the **Apply** and then **OK** button; the intensity vs height data is added to the **Intensity** page of the **Add New : Wind Definitions** dialog box, as shown in Figure 6-51.

Step 8: Choose the **Add** button; the **Exposures** page is displayed. Specify the value **0.75** in the **Factor** edit box and then choose the **Add** button.

Step 9: Again, specify the value **0.8** in the **Factor** edit box and choose the **Add** button to add the exposure factor. Next, close the **Add New : Wind Definitions** dialog box.

Step 10: Assign the exposure factor 0.75 to the nodes 1, 4, 6, 9, 12, and 14 and assign exposure factor 0.8 to the nodes 17, 20, and 22 using the **Use Cursor to Assign** method in the main window.

Step 11: Select the **Load Cases Details** node in the **Load & Definition** window and choose the **Add** button; the **Add New : Load Cases** dialog box is displayed.

Step 12: Enter the text **WL +X** in the **Title** text box and select **Wind** from the **Loading Type** drop-down list. Next, choose the **Add** button to add the load case and close the dialog box.

Step 13: Select the **WL +X** load case available under the **Load Cases Details** node and then choose the **Add** button; the **Add New : Load Items** dialog box is displayed.

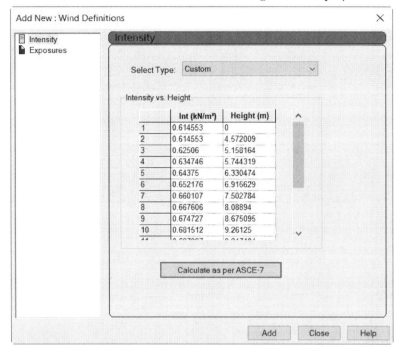

*Figure 6-51 Values specified in the **Intensity vs. Height** table in the **Intensity** page*

Step 14: Select the **Wind Load** option in the left pane of the dialog box; the **Wind Load** page is displayed in the right pane of the dialog box.

Step 15: In this page, ensure that **1 : Regular** is selected in the **Select Type** drop-down list and **X (Windward Face)** radio button is selected in the **Exposed Surface and Direction** area. Specify **1.1** in the **Factor** edit box.

Step 16: Choose the **Add** button; the wind load is added under the **WL +X** load case in the **Load & Definition** window. Next, close the **Add New : Load Items** dialog box.

Step 17: Select the added load under the **WL +X** load case; the wind load is applied as joint load on the left wall of the structure, as shown in Figure 6-52.

Step 18: Choose the **Save As** option from the **File** menu; the **Save As** dialog box is displayed. In this dialog box, specify the name *c06_staad_connect_ex5* in the **File name** edit box and save it at an appropriate location. Close the file by using the **Close** option in the **File** menu.

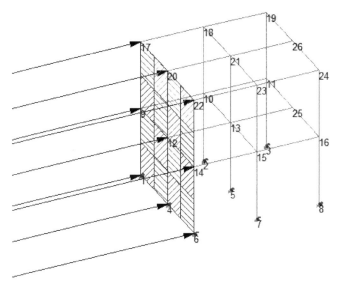

Figure 6-52 *Wind load applied on the left wall of the structure*

Snow Load

Snow load allows you to generate snow loading on the structure as per the code ASCE-7-2002. To generate snow load, select the **Snow Load** node in the left pane of the **Add New : Load Items** dialog box; the **Snow Load** page will be displayed in the right pane of the dialog box, refer to Figure 6-53. The options in this page are inactive. To activate these options, first you need to define the snow load, which is discussed next.

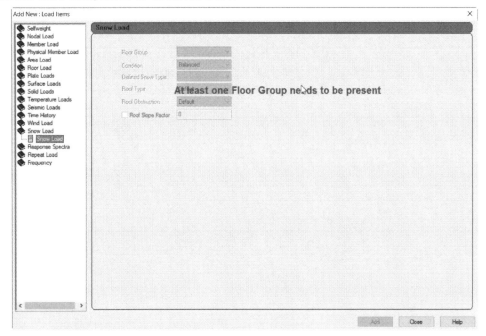

Figure 6-53 *The **Snow Load** page in the **Add New : Load Items** dialog box*

Defining Snow Load

To define snow load, expand the **Definition** node in the **Load & Definition** window and then select the **Snow Definition** option and choose the **Add** button; the **Add New : Snow Definition** dialog box will be displayed, as shown in Figure 6-54. The options in this dialog box are discussed next.

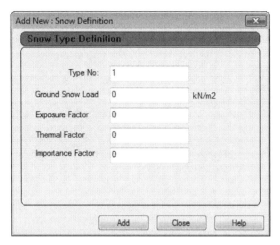

Figure 6-54 The **Add New : Snow Definition** *dialog box*

Type No

In the **Type No** edit box, you will specify the snow load type number.

Ground Snow Load

Specify a value in the **Ground Snow Load** edit box to calculate the design snow load.

Exposure Factor

Specify a value in the **Exposure Factor** edit box depending upon the type of exposure such as partially exposed, fully exposed, or sheltered and also the terrain category.

Thermal Factor

Specify a value in the **Thermal Factor** edit box depending upon the thermal condition of the structure.

Importance Factor

In the **Importance Factor** edit box, specify a value depending upon the structure category.

Next, choose the **Add** button to add the snow load type in the **Load & Definition** window and then close the **Add New : Snow Definition** dialog box.

Note
Before defining snow load as primary load, you need to create floor groups. The process to do so has been discussed earlier in this chapter.

After defining the snow load, you need to define it as primary load. To do so, select the required load case under the **Load Cases Details** in the **Load & Definition** window, and choose the **Add** button; the **Add New : Load Items** dialog box will be displayed. In the left pane of the dialog box, select the **Snow Load** option; the **Snow Load** page will be displayed in the right pane, refer to Figure 6-55. The options in this page are discussed next.

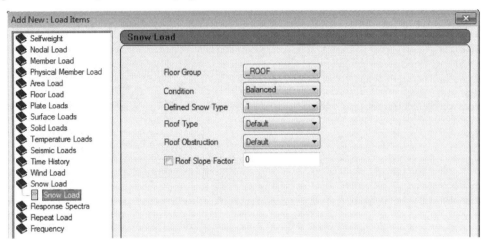

*Figure 6-55 The **Snow Load** page in the **Add New : Load Items** dialog box*

Select the floor group on which the snow load will be applied from the **Floor Group** drop-down list. Here, make sure that the members on which loading will be applied are defined as floor groups. Select the desired option from the **Condition** drop-down list. Select the previously defined load type number from the **Defined Snow Type** drop-down list. Specify the roof type such as mono, hipped, or gable by selecting an appropriate option from the **Roof Type** drop-down list. Select an option for the obstructed or unobstructed roof from the **Roof Obstruction** drop-down list. For sloped roofs, select the **Roof Slope Factor** check box and then specify the slope factor in the edit box next to the check box. Next, choose the **Add** button; the snow loads will be added under the selected load case and it will be automatically assigned to the structure.

Note
You can refer to ASCE 7-2002 code of standard for parameters required for defining the snow load.

Response Spectra

In the response spectrum method, you can analyze a structure dynamically. In this method, the dynamic response of a structure is calculated on the basis of the natural response of individual mode of vibration. The resultant is the summation of responses of each individual mode of vibration. To define response spectra loading, select the required load case in the **Load & Definition** window and then choose the **Add** button; the **Add New : Load Items** dialog box will be displayed. In this dialog box, select the **Response Spectra** node in the left pane; the **Response Spectrum** page will be displayed in the right pane of the dialog box, refer to Figure 6-56. The options displayed in this page are discussed next.

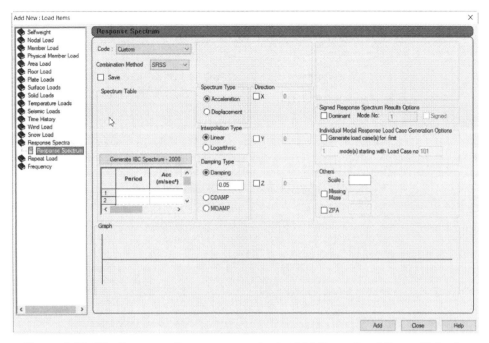

*Figure 6-56 The **Response Spectrum** page in the **Add New : Load Items** dialog box*

Select the required code from the **Code** drop-down list. Select the desired combination method from the **Combination Method** drop-down list. These combination methods are SRSS (Square root of summation of squares), ABS (Absolute sum method), CQC (Complete quadratic combination), TEN (Ten percent method), ASCE (as per ASCE-48 manual), and CSM (Closely spaced modes method). In the **Spectrum Type** area, select the **Acceleration** radio button for the period vs acceleration response spectrum curve or select the **Displacement** radio button for the period vs displacement curve. Specify the global direction in which spectrum will be applied by selecting **X**, **Y**, or **Z** check box and specify the factor applicable to each direction in their respective edit boxes. In the **Interpolation Type** area, specify the interpolation method to be used for the spectral value. The **Linear** radio button is selected by default. In the **Damping Type** area, specify the damping type to be used in the response spectrum analysis. In the **Others** area, specify a value in the **Scale** edit box. This value will get multiplied with the spectral data during the analysis. Select the **Missing Mass** check box to apply the missing mass correction. Select the **ZPA** check box to use the Zero Period Acceleration. Enter the period vs acceleration or period vs displacement value in the table. Note that the spectral data values should be in the increasing order of the period. As you provide the values, the curve will be displayed at the bottom in the dialog box. Next, choose the **Add** button to add the load in the **Load & Definition** window.

STAAD.Pro also allows you to automatically generate the response spectrum data as per IBC/ASCE code. To generate the data automatically, choose the **Generate IBC Spectrum - 2000** button in the **Response Spectrum** page of the **Add New : Load Items** dialog box; the **Spectrum Parameters::IBC 2000** dialog box will be displayed, as shown in Figure 6-57. The options in this dialog box are discussed next.

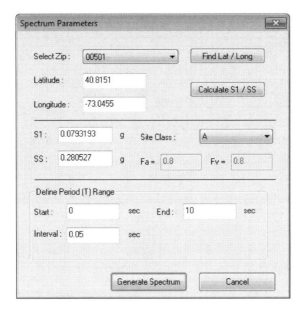

Figure 6-57 *The **Spectrum Parameters** dialog box*

Select the required zip code from the **Select Zip** drop-down list. On selecting the zip code, the latitudinal and longitudinal values get updated automatically in the **Latitude** and **Longitude** edit boxes. If you do not know the zip code, then enter the latitudinal and longitudinal values in their corresponding edit boxes. After specifying the values manually in the **Latitude** and **Longitude** edit boxes, choose the **Calculate S1/SS** button; the site coefficient values of S1 and SS will be filled automatically in the **S1** and **SS** edit boxes. If the zip code is known, then the site coefficient value gets filled automatically. Select the required site class based on the soil type from the **Site Class** drop-down list; the values in the F_a and F_v edit boxes will be set automatically.

In the **Define Period (T) Range** area, specify the start time of the period for creating the response spectrum in the **Start** edit box. Similarly, specify the end value of the time period in the **End** edit box. Specify the time interval at which the spectral data will be calculated in the **Interval** edit box.

After specifying all the parameters, choose the **Generate Spectrum** button; the spectrum based on the specified data will be displayed in the **Generated Spectrum** dialog box, as shown in Figure 6-58. Next, close the **Generated Spectrum** dialog box and then choose the **Add** button in the **Add New : Load Items** dialog box. The load will be added under the selected load case in the **Load & Definition** window and the added load will be automatically assigned to the structure. Close the **Add New : Load Items** dialog box.

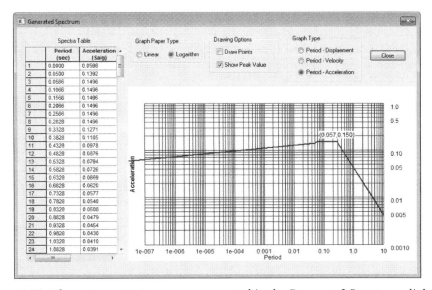

Figure 6-58 *The response spectrum curve generated in the* **Generated Spectrum** *dialog box*

Example 6

In this example, you will open the *c06_staad_connect_ex4.std* file. Next, you will define the response spectrum loading as per IBC 2006.

The following steps are required to complete this example:

Step 1: Open the file *c06_staad_connect_ex4.std* in STAAD.Pro; the model is displayed in the user interface screen, as shown in Figure 6-59.

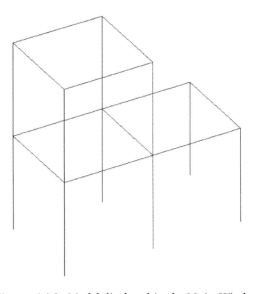

Figure 6-59 *Model displayed in the Main Window*

Step 2: Choose the **Loading** tab; the **Load & Definition** window is displayed. In the **Load & Definition** window; select the **Load Cases Details** and choose the **Add** button; the **Add New: Load Cases** dialog box is displayed. In this dialog box, specify the title **Response Spectra** in the **Title** text box and choose the **Add** button; the load case is added. Next, close the dialog box.

Step 3: Now, select the **Response Spectra** load case in the **Load & Definition** window and choose the **Add** button; the **Add New : Load Items** dialog box is displayed.

Step 4: In this dialog box, specify **1** in the **Factor** edit box and select the **X** radio button. Next, choose the **Add** button; the **SELFWEIGHT X 1** is added under the **Response Spectra** load case.

Step 5: Repeat the procedure followed in step 4 and add the self weight in **Y** and **Z** direction for factor 1 and close the **Add New : Load Items** dialog box.

Step 6: Next, select the **Floor Load** node in the left pane of the dialog box; the **Floor** page is displayed in the right pane of the dialog box. In this page, specify the values as given in Figure 6-60.

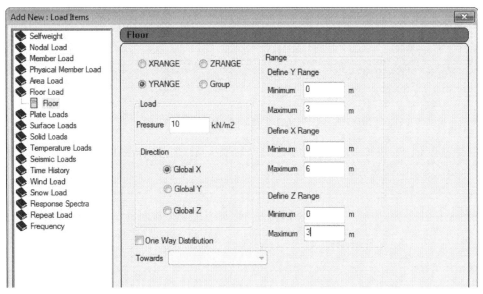

Figure 6-60 *Values specified for a floor load in the **Floor** page*

Step 7: Choose the **Add** button to add the floor load. Similarly, add the same load intensity in the **Y** and **Z** directions. You need to select the **Global Y** radio button for the second floor loading and the **Global Z** radio button for the third floor loading.

Step 8: Next, define a different load with the load intensity **8**, as shown in Figure 6-61.

Step 9: Choose the **Add** button to add the load. Similarly, add two more loads for **Y** and **Z** directions. Next, select the **Response Spectra** node in the left pane of the dialog box; the **Response Spectrum** page is displayed in the right pane of the dialog box.

Step 10: In this dialog box, specify the values, as shown in Figure 6-62.

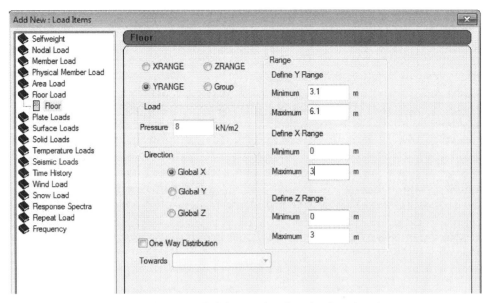

Figure 6-61 *Values specified for another floor load in the **Floor** page*

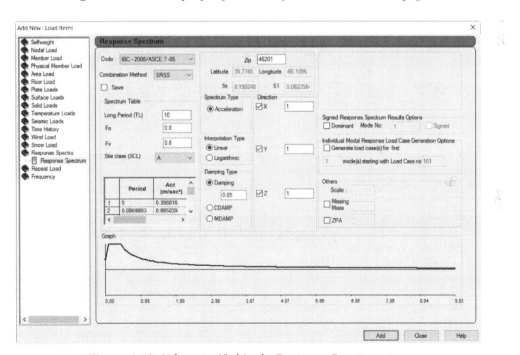

Figure 6-62 *Values specified in the **Response Spectrum** page*

Step 11: Choose the **Add** button to add the load and close the dialog box. Next, using the **Assign To View** method, assign the added self weights to the structure.

Step 12: Choose the **Editor** tool from the **Utilities** panel of the **Utilities** tab; the **Warning** message box is displayed. Choose the **Save** button; the **STAAD Editor** window is displayed. In this window, the command for the **Response Spectra** load case is displayed as given next.

```
LOAD 2 LOADTYPE None TITLE RESPONSE SPECTRA
SELFWEIGHT X 1
SELFWEIGHT Y 1
SELFWEIGHT Z 1
FLOOR LOAD
YRANGE 0 3 FLOAD 10 XRANGE 0 6 ZRANGE 0 3 GY
YRANGE 0 3 FLOAD 10 XRANGE 0 6 ZRANGE 0 3 GX
YRANGE 0 3 FLOAD 10 XRANGE 0 6 ZRANGE 0 3 GZ
YRANGE 3.1 6.1 FLOAD 8 XRANGE 0 3 ZRANGE 0 3 GX
YRANGE 3.1 6.1 FLOAD 8 XRANGE 0 3 ZRANGE 0 3 GY
YRANGE 3.1 6.1 FLOAD 8 XRANGE 0 3 ZRANGE 0 3 GZ
SPECTRUM SRSS IBC 2006 X 1 Y 1 Z 1 ACC DAMP 0.05 LIN
ZIP 46201 SITE CLASS A FA 0.800 FV 0.800 TL 10.000
```

Step 13: Choose the **Save As** option from the **File** menu; the **Save As** dialog box is displayed. In this dialog box, specify the name *c06_staad_connect_ex6* in the **File name** edit box and save it at an appropriate location. Close the file by choosing the **Close** option from the **File** menu.

Repeat Load

A repeat load is a primary load case created using a combination of previously defined primary load cases for which STAAD.Pro analyses the structure. To define this load type, select the **Repeat Load** node in the left pane of the **Add New : Load Items** dialog box; the **Repeat Load** page will be displayed in the right pane of the dialog box. In this page, all the primary load cases defined earlier will be displayed in the **Available Load Cases** list box. To include a primary load case, first select it from the list box and then choose the forward button; the load will be shifted to the **Repeat Load Definition** list box, refer to Figure 6-63. Next, specify a multiplying factor in the **Factor** edit box in the **Repeat Load Definition** list box. You can use the CTRL key and select as many load cases as you want. To shift all the load cases at a time, choose the [**>>**] button. Next, choose the **Add** button to add the load case. Two more options are displayed under the **Repeat Load** node in the left pane of the **Add New : Load Items** dialog box. These options are : **Reference Load** and **Notional Load**. The **Reference Load** has been discussed earlier in this chapter.

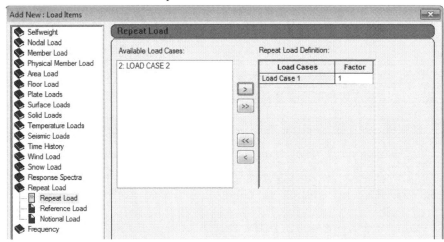

Figure 6-63 *The **Repeat Load** page in the **Add New : Load Items** dialog box*

The **Notional load** is defined as a percentage of gravity loads. In this load type, the primary and reference load cases can be selected from the **Primary Load Cases** and **Reference Load Cases** area and a percentage of them will get applied to the structure. To add this load type, select it; the **Notional Load** page will be displayed, as shown in Figure 6-64. Next, shift the required primary and reference load cases to the **Notional Load Definition** list box. Specify the required factors in the **Factor** column and choose the **Add** button to add the load case in the **Load & Definition** window.

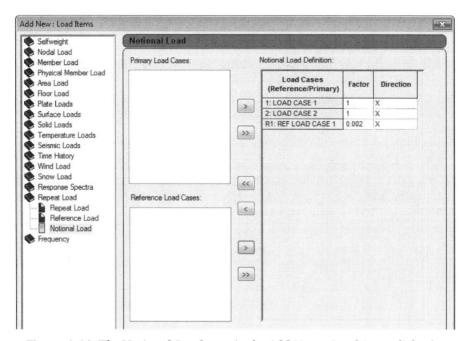

Figure 6-64 *The **Notional Load** page in the **Add New : Load Items** dialog box*

Frequency

There are two methods in STAAD.Pro used for calculating the frequencies of a structure. These methods are: Rayleigh method and EigenValue Extraction method.

To use Rayleigh Frequency method, a primary load case should be created first and then the **Rayleigh Frequency** load option should be added to the created load case. To add Rayleigh Frequency, select the **Frequency** node in the left pane of the **Add New : Load Items** dialog box; the **Rayleigh Frequency** page will be displayed in the right pane of the dialog box. Choose the **Add** button to add the **CALCULATE RAYLEIGH FREQUENCY** command under the selected load case in the **Load & Definition** window.

EigenValue Extraction method is used to calculate relevant frequencies and mode shapes. To use this method, select the **Modal Calculation** option available under the **Frequency** node in the left pane of the **Add New : Load Items** dialog box; the **Modal Calculation** page will be displayed in the right pane of the dialog box. Next, choose the **Add** button; the **MODAL CALCULATION REQUESTED** command will be added under the selected load case in the **Load & Definition** window.

LOAD GENERATION

The **Load Generation** option is used to create primary load case using the data of pre defined vehicle load definitions. To use this option, first select the **Load Cases Details** node in the **Load & Definition** window and then choose the **Add** button; the **Add New : Load Cases** dialog box will be displayed. In this dialog box, select the **Load Generation** node in the left pane of the dialog box; the **Load Generation** page will be displayed in the right pane of the dialog box. In this page, you can define the number of loads to be generated only after defining the vehicle load which is discussed next.

Defining Vehicle Loading

In STAAD.Pro, different types of moving loads can be defined. To define a vehicle load, expand the **Definitions** node in the **Load & Definition** window and select the **Vehicle Definitions** option. Next, choose the **Add** button; the **Add New : Vehicle Definitions** dialog box will be displayed, as shown in Figure 6-65. In this dialog box, you can use any of the three nodes to define the moving load. These nodes are discussed next.

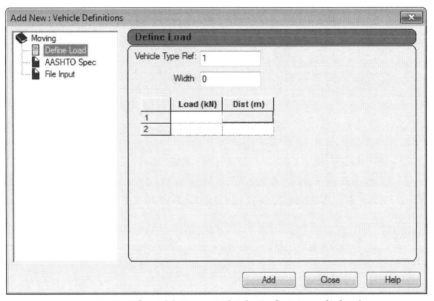

*Figure 6-65 The **Add New : Vehicle Definitions** dialog box*

Define Load

The **Define Load** option is selected by default in the left pane of the dialog box and the **Define Load** page will be displayed in the right pane of the dialog box, refer to Figure 6-65. In the **Vehicle Type Ref** edit box of this page, the reference number will be specified by default. Specify the spacing between the wheels in the **Width** edit box. Here, width is the distance between the parallel wheels. Next, you have to specify the value of the concentrated loads acting on the wheels and the distance between them in the **Load (kN)** and **Dist (m)** columns. After specifying the values, choose the **Add** button; the vehicle type will be defined and added under the **Vehicle Definitions** in the **Load & Definition** window.

AASHTO Spec

The vehicle loading can also be defined using the AASHTO specifications, which are available in the **AASHTO Spec** page. To access this page, select the **AASHTO Spec** option in the left pane of the dialog box; the **AASHTO Spec** page will be displayed in the right pane, as shown in Figure 6-66.

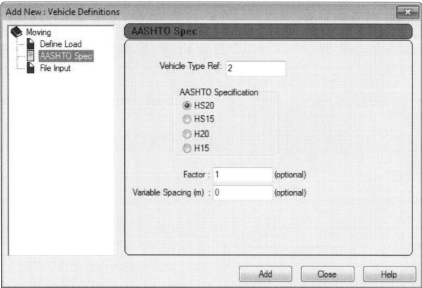

Figure 6-66 The **AASHTO Spec** page in the **Add New : Vehicle Definitions** dialog box

The vehicle type reference number will be specified by default in the **Vehicle Type Ref** edit box. In the **AASHTO Specification** area, specify the type of loading by selecting the **HS20**, **HS15**, **H20**, or **H15** radio button. These are the series of truck loadings as per the AASHTO (American Association of State and Highway Transportation Officials) specifications. Here, HS20 means a 20 ton heavy semi-trailer truck and H20 means a 20 ton heavy truck. Next, specify the multiplying factor in the **Factor** edit box. You can also use the default value, which is **1**. In the **Variable Spacing (m)** edit box, specify the wheel spacing. It is the spacing between the loads perpendicular to the direction of movement. Next, choose the **Add** button to add the vehicle type in the **Load & Definitions** window.

File Input

Select the **File Input** option from the left pane of the dialog box; the **File Input** page will be displayed in the right pane, refer to Figure 6-67. In this page, you can import the moving load data from an external file. To do so, choose the **Import** button; the **Open** dialog box will be displayed. In this dialog box, browse to the file location and select the file. Next, choose the **Open** button; the file will be loaded and its name will be displayed in the **File Name** text box. The load name will be filled automatically in the **Load Name** drop-down list. Specify the multiplying factor in the **Factor** edit box. Next, choose the **Add** button; the load will be added under the **Vehicle Definitions** node in the **Load & Definition** window. Next, close the **Add New : Vehicle Definitions** dialog box.

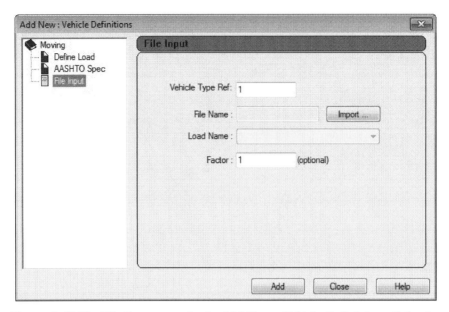

*Figure 6-67 The **File Input** page in the **Add New : Vehicle Definitions** dialog box*

After defining the vehicle type, go to the **Load Generation** page in the **Add New : Load Cases** dialog box, as discussed earlier. Figure 6-68 shows the **Load Generation** page in the **Add New : Load Cases** dialog box.

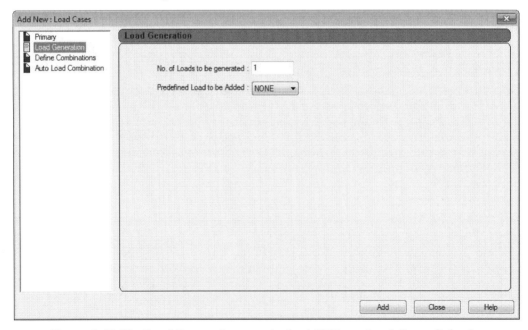

*Figure 6-68 The **Load Generation** page in the **Add New : Load Cases** dialog box*

In the **Load Generation** page, specify the number of load cases to be generated in the **No. of Loads to be generated** edit box. The number of load cases represent the number of positions for the vehicle on the structure. You can also include any of the predefined load cases in the moving load. The predefined load can be the self weight of the structure or the dead load. To include the predefined load, select the required load case from the **Predefined Load to be Added** drop-down list. Next, choose the **Add** button; the **GENERATE LOAD** sub node will be added under the **Load Cases Details** in the **Load & Definition** window. Now, select this added sub-node and choose the **Add** button; the **Load Generation Type** page will be displayed in the **Add New : Load Cases** dialog box, refer to Figure 6-69. The options displayed in this page are discussed next.

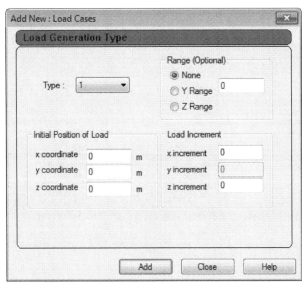

*Figure 6-69 The **Load Generation Type** page in the **Add New : Load Cases** dialog box*

In the **Type** drop-down list, select the vehicle type number which has been defined earlier. In the **Initial Position of Load** area, specify the location of the vehicle. Specify the required values in the **x coordinate**, **y coordinate**, and **z coordinate** edit boxes. In the **Range (Optional)** area, select the required radio button to specify a range. In the **Load Increment** area, specify the required values in the **x increment** and **z increment** edit boxes. For example, if you will specify **3ft** in the **z increment** edit box then the vehicle will move 3ft at a time. The vehicle will only move in the horizontal plane so the **y increment** edit box will be in the inactive mode. Next, choose the **Add** button; the vehicle position will get defined and added under the **GENERATE LOAD** sub node in the **Load & Definition** window and then close the **Add New : Load Cases** dialog box.

Example 7

In this example, first you will open the *c06_staad_connect_ex7_start.std* file. Next, you will define vehicle loading as per AASHTO specifications and then define its position on a bridge deck.

Steps required to complete this example are given next:

Step 1: Open the file *c06_staad_connect_ex7_start.std* in STAAD.Pro; the model is displayed in the main window, as shown in Figure 6-70.

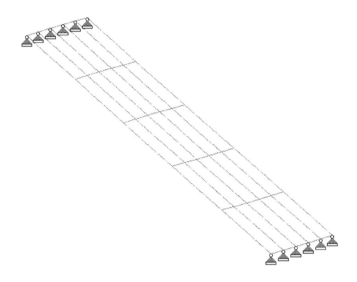

Figure 6-70 *The Bridge Deck model*

Step 2: Choose the **Loading** tab; the **Load & Definition** window is displayed. In this window, expand the **Definitions** node and select the **Vehicle Definitions** sub node.

Step 3: Next, choose the **Add** button; the **Add New : Vehicle Definitions** dialog box is displayed. In this dialog box, select the **AASHTO Spec** option; the **AASHTO Spec** page is displayed in the right pane of the dialog box.

Step 4: In this page, specify the values for the vehicle type, as shown in Figure 6-71. Next, choose the **Add** button; the **TYPE 1 HS20 1 4.2** type is defined and added under the **Vehicle Definitions** sub node. Now, close the **Add New : Vehicle Definitions** dialog box.

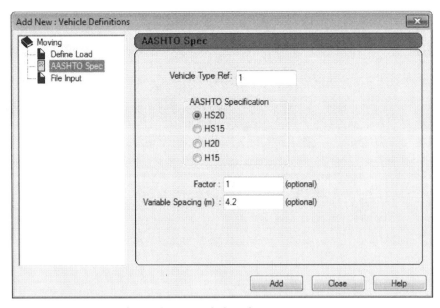

Figure 6-71 *Values specified in the **AASHTO Spec** page*

Step 5: Next, select the **Load Cases Details** node in the **Load & Definition** window and then choose the **Add** button; the **Add New : Load Cases** dialog box is displayed.

Step 6: In this dialog box, select the **Load Generation** node in the left pane; the **Load Generation** page is displayed in the right pane.

Step 7: In this page, specify **10** in the **No. of Loads to be generated** edit box and select **1** in the **Predefined Loads to be Added** drop-down list. Next, choose the **Add** button; the **GENERATE LOAD, ADD LOAD 1** is added under the **Load Cases Details** node. Now, close the **Add New : Load Cases** dialog box.

Step 8: Select the **GENERATE LOAD, ADD LOAD 1** in the **Load & Definition** window and choose the **Add** button; the **Add New : Load Cases** dialog box is displayed.

Step 9: In this dialog box, specify the values for the vehicle position, as shown in Figure 6-72. Next, choose the **Add** button; the position will be defined and added under the **GENERATE LOAD, ADD LOAD 1** in the **Load & Definition** window. Close the **Add New : Load Cases** dialog box.

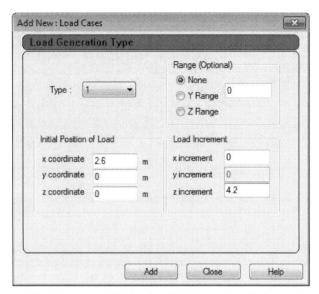

*Figure 6-72 Values specified in the **Add New : Load Cases** dialog box*

Step 10: Now, click on the defined position available under the **GENERATE LOAD, ADD LOAD 1** in the **Load & Definition** window; the vehicle is displayed and placed on the bridge, as shown in Figure 6-73.

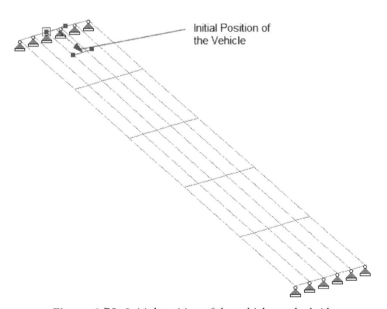

Figure 6-73 Initial position of the vehicle on the bridge

Step 11: Choose the **Editor** tool from the **Utilities** panel of the **Utilities** tab; the **Warning** message box is displayed. Choose the **Save** button in this message box; the **STAAD Editor** window is displayed. In this window, the command for the vehicle load case is displayed as given next:

```
DEFINE MOVING LOAD
TYPE 1 HS20 1 4.2
LOAD 1 LOADTYPE None TITLE LOAD CASE 1
SELFWEIGHT Y -1 LIST 1 TO 27 33 TO 43 49 TO 59 65 TO 75
LOAD GENERATION 9 ADD LOAD 1
TYPE 1 2.6 0 0 ZINC 4.2
```

Step 12: Choose the **Save As** option from the **File** menu; the **Save As** dialog box is displayed. In this dialog box, specify the name *c06_staad_connect_ex7* in the **File name** edit box and save it at an appropriate location. Close the file by choosing the **Close** option from the **File** menu.

DEFINING LOAD COMBINATIONS

In STAAD.Pro, combination loading is defined to combine the result of analysis performed for different primary load cases. A combination load sums up the analysis results of certain individual load cases. This summation is suitable for linear analysis in which the results of analysis are superimposed. To define combination loading, select the **Load Cases Details** node in the **Load & Definition** window and then choose the **Add** button; the **Add New : Load Cases** dialog box will be displayed. In this dialog box, select the **Define Combinations** node in the left pane of the dialog box; the **Define Combinations** page will be displayed in the right pane, refer to Figure 6-74. The options displayed in this page are discussed next.

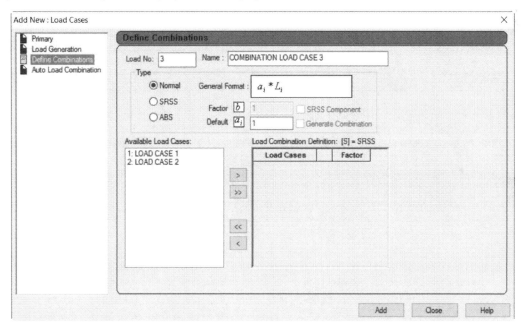

*Figure 6-74 The **Define Combinations** page in the **Add New : Load Cases** dialog box*

The load case number will be automatically filled in the **Load No** edit box. Specify the desired name in the **Name** text box. In the **Type** area, three radio buttons are available for three combination methods: Normal, SRSS, and ABS that are discusses next.

The **Normal** radio button is selected by default. This method uses the combination a_i*L_i. Here, a_i denotes the multiplying factor and L_i denotes the load case number. Now, to define this loading, select the load cases individually in the **Available Load Cases** area and move them to the **Load Combination Definition** area using the single arrow button. Next, specify their multiplying factors in the **Factor** column. Choose the **Add** button to add the combination loading in the **Load & Definition** window.

The SRSS (Square root sum of squares) method is required for nuclear codes such as ASCE 4-98. To use this method, select the **SRSS** radio button; the combination expression will be displayed in the **General Format** box. Now, select the required load cases from the **Available Load Cases** area and move them to the **Load Combination Definition** area using the single arrow button. Next, specify their multiplying factors in the **Factor** column.

To use the ABS (Absolute) combination method, select the **ABS** radio button; the combination expression will be displayed in the **General Format** box. The method for defining this combination is same as discussed above. Choose the **Add** button to add the required combination loading in the **Load & Definition** window. Next, close the **Add New : Load Cases** dialog box.

DEFINING LOAD COMBINATIONS AUTOMATICALLY

STAAD.Pro automatically creates load combinations based on the standard load combination factors using the codes like ACI, AISC, UBC, IBC, Indian Code, British, and NBCC 1995. To define auto load combinations, first ensure that the primary loads have been created. Next, select the **Load Cases Details** in the **Load & Definition** window and choose the **Add** button; the **Add New : Load Cases** dialog box will be displayed. In this dialog box, select the **Auto Load Combination** option from the left pane; the **Auto Load Combination** page will be displayed in the right pane, refer to Figure 6-75. The options displayed in this page are discussed next.

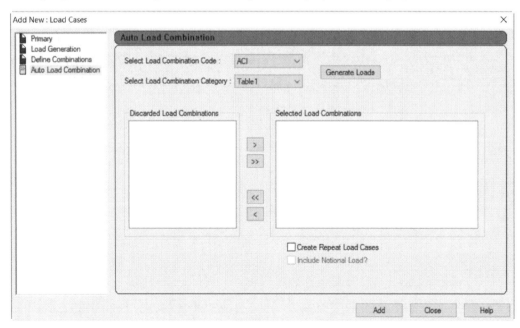

*Figure 6-75 The **Auto Load Combination** page in the **Add New : Load Cases** dialog box*

Note

While creating primary load cases, do not forget to define load category such as wind, seismic, and so on.

Select the required combination code from the **Select Load Combination Code** drop-down list. Next, select the required load combination category from the **Select Load Combination Category** drop-down list. Now, choose the **Generate Loads** button; various combinations will be generated and displayed in the **Selected Load Combinations** list box, as shown in Figure 6-76.

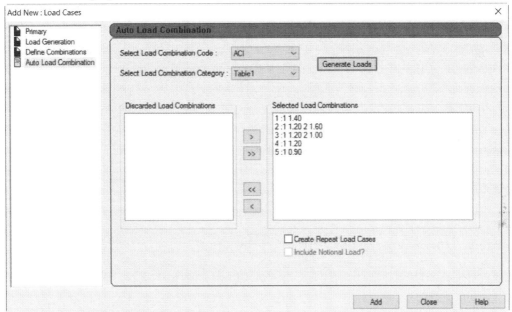

*Figure 6-76 Load combinations generated in the **Selected Load Combinations** list box*

To exclude any of the combinations, select it and then move it to the **Discarded Load Combinations** list box using the arrow button. Next, choose the **Add** button; all the combinations will be generated and added under the **Load Cases Details** in the **Load & Definition** window.

Self-Evaluation Test

Answer the following questions and compare them to those given at the end of this chapter:

1. Nodal loads are applied at the _____ in a structure.

2. _____ loads act at a point in the structure.

3. The_____ option is used to define uniform pressure on surface elements.

4. The_____ option is used to define varying pressure on each joint in a plate element.

5. The loading effect in poststress load gets transmitted to the connected structural members. (T/F)

6. You cannot define the pressure acting partially on a surface element. (T/F)

7. Before creating the seismic load, you first need to define the seismic parameters. (T/F)

Review Questions

Answer the following questions:

1. Which of the following options is used to define temperature loading?

    (a) Temperature Load   (b) Solid Load
    (c) Wind Load   (d) None of these

2. Which of the following loadings is defined to perform dynamic analysis?

    (a) Vehicle   (b) Temperature
    (c) Snow   (d) Seismic

3. Which of the following loading options varies with time?

    (a) Nodal Load   (b) Time History
    (c) Surface Load   (d) None of these

4. Which of the following options is used to define snow loading?

    (a) Plate Load   (b) Member Load
    (c) Snow Load   (d) Wind Load

5. In STAAD.Pro, wind load is applied as nodal load in a structure. (T/F)

6. The repeat load combines the results of analysis performed for different primary load cases. (T/F)

7. The Auto Load Combination option creates load combinations automatically. (T/F)

Answers to Self-Evaluation Test

1. joints, **2.** Concentrated, **3.** Pressure on Full Surface, **4.** Element Joint Load, **5.** F, **6.** F, **7.** T

Chapter 7

Performing Analysis, Viewing Results, and Preparing Report

Learning Objectives

After completing this chapter, you will be able to:
- *Apply various pre analysis print commands*
- *Apply various perform analysis commands*
- *Apply various post analysis print commands*
- *View the results*
- *Create a customized report*

INTRODUCTION

In the previous chapter, you learned to assign different types of loads to a structure. In this chapter, you will learn to perform different types of analysis to view the structural behavior after application of these loads. After the analysis, you will view the results graphically and numerically for support reactions, member forces, bending moments, deflection, and so on. Later on, you will prepare a report which will include the graphical and numerical results, structural model, properties, and loading information.

In this chapter, all the above steps are categorized into five groups: Pre Analysis Print, Performing Analysis, Post Analysis Print, Viewing Results, and Preparing Report. These steps are discussed in detail in this chapter.

PRE ANALYSIS PRINT

In STAAD.Pro, after assigning loads and other properties, you will specify the commands for displaying some specific information about the structure in the STAAD Output file. These commands are known as Pre Print commands. To specify Pre Print commands, choose the **Pre Analysis Commands** button from the **Analysis Data** panel of the **Analysis and Design** tab; the **Pre Analysis Print - Whole Structure** window will be displayed. In this window, choose the **Define Commands** button; the **Analysis/Print Commands** dialog box will be displayed, as shown in Figure 7-1. This dialog box comprises of several tabs which are used for adding Pre-Print commands. These tabs are discussed next.

Figure 7-1 The Analysis/Print Commands dialog box

Problem Statistics

The **Problem Statistics** tab is chosen by default. In this tab, choose the **Add** button; the **PRINT PROBLEM STATISTICS** command will be added to the **Pre Analysis Print - Whole Structure** window. This command is used to print the structural information such as total number of nodes, members, supports, disk space requirement, and other information of the output file.

Joint Coordinates

You can print the joint coordinate values in the output file. To do so, choose the **Joint Coordinates** tab in the **Analysis/Print Commands** dialog box. Next, choose the **Add** button; the **PRINT JOINT COORDINATES** command will be added to the **Pre Analysis Print - Whole Structure** window. You can see a question mark displayed before the **PRINT JOINT COORDINATES** command in the **Pre Analysis Print - Whole Structure** window. This shows that the command has not been assigned to the structure. You can assign the command by using any of the four assignment methods which have been discussed earlier.

Member Information

Member information such as member length, member incidences, beta angles, member specifications, and so on can be printed in the output files. To do so, choose the **Member Information** tab and then choose the **Add** button; the **PRINT MEMBER INFORMATION** command will be added to the **Pre Analysis Print - Whole Structure** window.

Material Properties

You can also print the material properties such as Young's Modulus, Shear Modulus, and so on in the output file. To do so, choose the **Material Properties** tab and then choose the **Add** button; the **PRINT MATERIAL PROPERTIES** command will be added to the **Pre Analysis Print - Whole Structure** window.

Support Information

You can print the information regarding the supports such as restraints, releases, and spring constant values in the output file. To do so, choose the **Support Information** tab and then choose the **Add** button; the **PRINT SUPPORT INFORMATION** command will be added in the **Pre Analysis Print - Whole Structure** window.

Member Properties

Member properties such as cross-sectional area, moment of inertia, and section modulus can also be printed in the output file. To do so, choose the **Member Properties** tab. Next, choose the **Add** button; the **PRINT MEMBER PROPERTIES** command will be added to the **Pre Analysis Print - Whole Structure** window.

Element Information

You can print element related information such as element incidences, thickness of element, and Poisson ratio in the output file. To do so, choose the **Element Information** tab and then choose the **Add** button; the **PRINT ELEMENT INFORMATION** command will be added to the **Pre Analysis Print - Whole Structure** window.

Solid Information

You can also add solid information such as element incidences and Poisson ratios for solid elements in the output file. To do so, choose the **Solid Information** tab and then choose the **Add** button; the **PRINT ELEMENT INFORMATION SOLID** command will be added to the **Pre Analysis Print - Whole Structure** window.

All

You can also print joint coordinates, member information, member properties, material properties, and support information together in the output file. To do so, choose the **All** tab and then choose the **Add** button; the **PRINT ALL** command will be added to the **Pre Analysis Print - Whole Structure** window.

PERFORMING ANALYSIS

In STAAD.Pro, after specifying the pre print commands, you have to specify the commands for the type of analysis to be performed. To perform an analysis, choose the **Analysis Commands** button from the **Analysis Data** panel of the **Analysis and Design** tab; the **Analysis/Print Commands** dialog box will be displayed, as shown in Figure 7-2. In this dialog box, different types of analysis are available as tabs. These tabs are: **Perform Analysis**, **PDelta Analysis**, **Perform Cable Analysis**, **Perform Direct Analysis**, **Perform Imperfection Analysis**, **Perform Buckling Analysis**, and **Perform Pushover Analysis, Change, Generate Floor Spectrum and Non Linear Analysis.** Some of these tabs are discussed next.

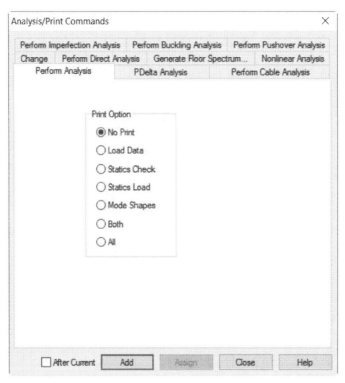

Figure 7-2 The **Analysis/Print Commands** dialog box with the **Perform Analysis** tab chosen

Perform Analysis

Using the **Perform Analysis** tab, you can add the **PERFORM ANALYSIS** command to the **Analysis - Whole Structure** window which will direct the program to proceed with the analysis. In the **Print Option** area of this tab, different print options are available which allow you to specify the analysis related data to be printed in the output file. These options are discussed next.

Select the **Load Data** radio button to print the entire load data in the output file. Select the **Statics Check** radio button to print the summation of the applied loads and support reactions along with the summation of moments of the loads and reactions taken around the origin in the output file. Select the **Statics Load** radio button to print the statics check results along with the summation of all the internal and external forces at each joint. Select the **Mode Shapes** radio button to print the mode shape values in the output file. Select the **Both** radio button to print both the statics check and load data results in the output file. Select the **All** radio button to print both the statics load and load data results in the output file. Next, choose the **Add** button; the **PERFORM ANALYSIS** command will be added to the **Analysis - Whole Structure** window. You can choose the **Close** button to close the **Analysis/Print Commands** dialog box.

PDelta Analysis

Using the **PDelta Analysis** tab, you can add the **PDELTA ANALYSIS** command to the **Analysis - Whole Structure** window. PDelta analysis is used for the structures subjected to lateral loads such as seismic loads. In this type of analysis, multiple iterations are required to produce accurate results. In this tab, specify the required number of iterations in the **Number of Iterations** edit box, refer to Figure 7-3. Alternatively, select the **Converge** check box and specify the number of iterations in the corresponding edit box. To include the geometric stiffness in the analysis, select the **Use Geometric Stiffness (Kg)** check box. To include the small delta effect, select the **Small Delta** check box. Select the required option from the **Print Option** area. Choose the **Add** button; the **PDELTA ANALYSIS** command will be added to the **Analysis - Whole Structure** window.

Perform Cable Analysis

The **PERFORM CABLE ANALYSIS** command is used to perform analysis for the cable members. In the **Perform Cable Analysis** tab, specify the maximum number of iterations per step in the **Eq-Iterations** edit box. Specify the convergence tolerance value in the **Eq-tolerance** edit box, refer to Figure 7-4. Specify the artificial stabilizing stiffness value in the **Stability Stiffness** edit box. Specify the minimum sag value in the **Sag Minimum** edit box. Specify the number of load steps to be applied in the **Load Steps** edit box. Specify the minimum amount of stiffness remaining after a cable sags in the **KSMALL** edit box. In the **Print Option** area, select the required option. Next, choose the **Add** button; the **PERFORM CABLE ANALYSIS** command will be added to the **Analysis - Whole Structure** window.

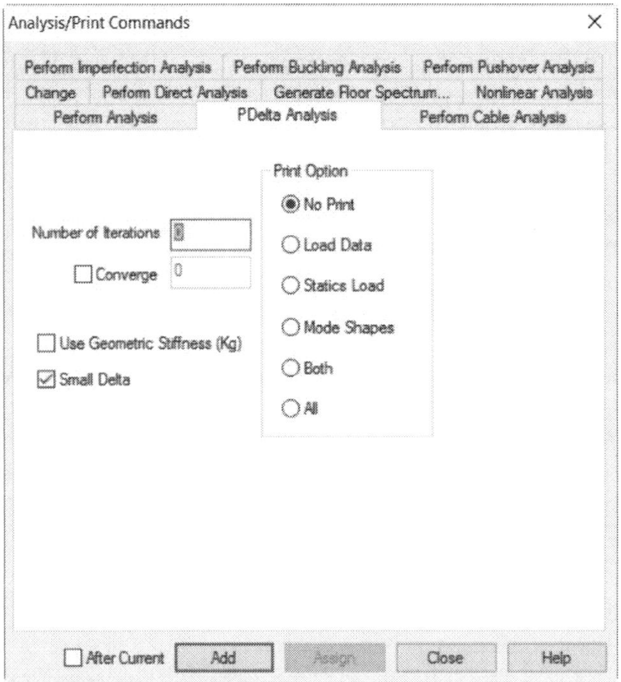

Figure 7-3 *The* **PDelta Analysis** *tab chosen in the* **Analysis/ Print Commands** *dialog box*

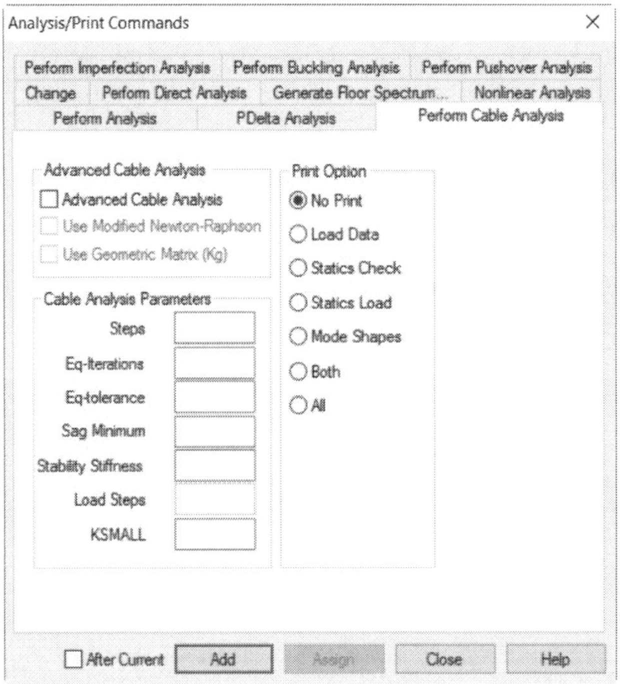

Figure 7-4 *The* **Perform Cable Analysis** *tab chosen in the* **Analysis/Print Commands** *dialog box*

Perform Direct Analysis

This analysis is a non-linear iterative analysis in which stiffness of the members is dependent upon the forces generated by the load. To perform direct analysis, first you need to define the **Direct Analysis Definition** in the **Load & Definition** window, as discussed in the previous chapters. After defining the **Direct Analysis Definition**, you will add the **PERFORM DIRECT ANALYSIS** command. To do so, choose the **Perform Direct Analysis** tab in the **Analysis/Print Commands** dialog box, refer to Figure 7-5. The options displayed in this tab are discussed next.

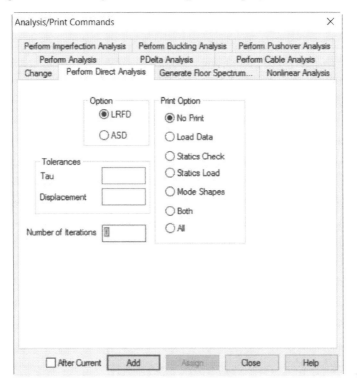

Figure 7-5 The **Perform Direct Analysis** *tab chosen in the* **Analysis/Print Commands** *dialog box*

In the **Option** area, select the required code by selecting the **LRFD** or **ASD** radio buttons. In the **Tolerances** area, specify the tau tolerance and displacement tolerance values in their respective edit boxes. Specify the required number of iterations for the analysis in the **Number of Iterations** edit box. Select the required option from the **Print Option** area. Next, choose the **Add** button; the **PERFORM DIRECT ANALYSIS** command will be added to the **Analysis - Whole Structure** window.

Perform Imperfection Analysis

Imperfection analysis is performed for the structural members that experience secondary forces due to imperfections in their geometry. The geometry imperfections are due to the curvature imperfection in the columns and beams. To perform the imperfection analysis, choose the **Perform Imperfection Analysi**s tab in the **Analysis/Print Commands** dialog box. In this tab, select the required print option from the **Print Option** area and then choose the **Add** button; the **PERFORM IMPERFECTION ANALYSIS** command will be added to the **Analysis - Whole Structure** window.

Perform Buckling Analysis

While performing the buckling analysis, STAAD.Pro performs the PDelta analysis including the geometric stiffness of plates and members due to large and small PDelta effects. To add **PERFORM BUCKLING ANALYSIS** command, choose the **Perform Buckling Analysis** tab in the **Analysis/Print Commands** dialog box. In this tab, specify the number of iterations in the **Number of Iterations** edit box. Next, select the required print option from the **Print Option** area and then choose the **Add** button; the **PERFORM BUCKLING ANALYSIS** command will be added to the **Analysis - Whole Structure** window.

Perform Pushover Analysis

The pushover analysis is performed by subjecting a structure to a consistently increasing pattern of lateral loads. Pushover analysis allows you to determine the non-linear force-displacement relationship. To perform this analysis, choose the **Perform Pushover Analysis** tab in the **Analysis/Print Commands** dialog box. In this tab, choose the **Add** button; the **PERFORM PUSHOVER ANALYSIS** command will be added to the **Analysis - Whole Structure** window. Next, close the **Analysis/Print Commands** dialog box.

POST ANALYSIS PRINT

After specifying the pre print and perform analysis commands, the next step is to add the post analysis print commands. Post analysis print commands are those commands which are used to print various analysis results in the output file. The analysis results include support reactions, member forces, stresses, displacements, and so on. To add post analysis print commands, choose the **Post Analysis Commands** button from the **Analysis Data** panel of the **Analysis and Design** tab; the **Post Analysis Print - Whole Structure** window will be displayed in the right pane of the interface. Next, choose the **Define Commands** button; the **Analysis/Print Commands** dialog box will be displayed, as shown in Figure 7-6. This dialog box comprises of several tabs using which you can add various post analysis print commands. Some of these tabs are discussed next.

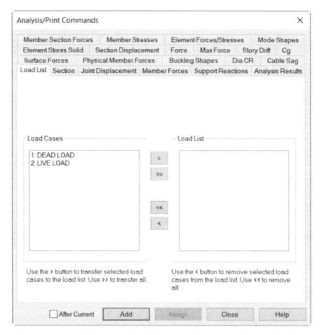

Figure 7-6 *The **Analysis/Print Commands** dialog box*

Load List

In the **Analysis/Print Commands** dialog box, the **Load List** tab is chosen by default. The **Load List** command is used to activate some of the load cases. By specifying the **Load List** command, the analysis will be performed only for the activated load cases. To add this command, first ensure that the **Load List** tab is chosen. Next, select the required load cases from the **Load Cases** area and then shift them to the **Load List** area. Next, choose the **Add** button; the **LOAD LIST** command will be added to the **Post Analysis Print - Whole Structure** window.

Joint Displacement

The joint displacements for all the specified joints of all the specified load cases can be printed in the STAAD output file. To do so, choose the **Joint Displacement** tab from the **Analysis/Print Commands** dialog box and then choose the **Add** button; the **PRINT JOINT DISPLACEMENTS** command will be added to the **Post Analysis Print - Whole Structure** window.

Member Forces

Member forces such as axial force, shear force in local Y and Z axes, torsional moment, and moments about local Y and Z axes can be calculated and printed in a tabular format in the STAAD output file. To do so, choose the **Member Forces** tab from the **Analysis/Print Commands** dialog box and then choose the **Add** button; the **PRINT MEMBER FORCES** command will be added to the **Post Analysis Print - Whole Structure** window. Next, assign this command to the required members in the structure.

Support Reactions

Support reactions such as F_x, F_y, F_z, M_x, M_y, and M_z can be calculated and printed in a tabular format in the output file. To do so, choose the **Support Reactions** tab from the **Analysis/Print**

Commands dialog box and then choose the **Add** button; the **PRINT SUPPORT REACTIONS** command will be added to the **Post Analysis Print - Whole Structure** window. Next, assign this command to the required joints in the structure.

Story Drift

You can also print the average lateral displacement of all the joints at each vertical level of the structure in the output file. To do so, choose the **Story Drift** tab from the **Analysis/Print Commands** dialog box and then choose the **Add** button; the **PRINT STORY DRIFT** command will be added to the **Post Analysis Print - Whole Structure** window.

Cg

You can print the coordinates of the point where centre of gravity is experienced on the structure in the output file. To do so, choose the **Cg** tab from the **Analysis/Print Commands** dialog box and then choose the **Add** button; the **PRINT CG** command will be added to the **Post Analysis Print - Whole Structure** window.

Mode Shapes

You can also print the mode shape values at every joint for all the calculated modes. To print mode shape values in the output file, choose the **Mode Shapes** tab from the **Analysis/Print Commands** dialog box and then choose the **Add** button; the **PRINT MODE SHAPES** command will be added to the **Post Analysis Print - Whole Structure** window.

Section Displacement

STAAD.Pro also calculates the section displacement and prints it in the output file. To do so, choose the **Section Displacement** tab from the **Analysis/Print Commands** dialog box and then choose the **Add** button; the **PRINT SECTION DISPL** command will be added to the **Post Analysis Print - Whole Structure** window.

Force

You can also print the maximum and minimum force/moment envelope values for every section in each specified load case. To do so, choose the **Force** tab from the **Analysis/Print Commands** dialog box. Next, specify the number of sections in the **Number of Sections** edit box. Choose the **Add** button; the **PRINT FORCE ENVELOPE** command will be added to the **Post Analysis Print - Whole Structure** window. Next, assign this command to the required members in the structure.

Analysis Results

If you want to print the joint displacements, support reactions, and member forces for all the specified joints/members with the load cases then choose the **Analysis Results** tab from the **Analysis/Print Commands** dialog box. Next, choose the **Add** button; the **PRINT ANALYSIS RESULTS** command will be added to the **Post Analysis Print - Whole Structure** window. Next, assign this command to the required members in the structure.

Member Stresses

You can also print member stresses in the output file. Member stresses include axial (axial force over area), bending-y, bending-z, shear stress in both local Y and Z directions, and combined stresses. To print member stresses, choose the **Member Stresses** tab from the **Analysis/Print Commands** dialog box and then choose the **Add** button; the **PRINT MEMBER STRESSES** command will be added to the **Post Analysis Print - Whole Structure** window. Next, assign this command to the required members in the structure.

Element Forces/Stresses

The element forces/stresses can also be printed in the output file. To do so, choose the **Element Forces/Stresses** tab. Next, select the required radio button and then choose the **Add** button; the **PRINT ELEMENT** command will be added to the **Post Analysis Print - Whole Structure** window. Next, assign this command to the required elements in the structure.

VIEWING RESULTS

After specifying the required commands, you need to perform the analysis to view the results. To do so, choose the **Run Analysis** button from the **Analysis** panel of the **Analysis and Design** tab; the **STAAD Analysis and Design** window will be displayed, as shown in Figure 7-7.

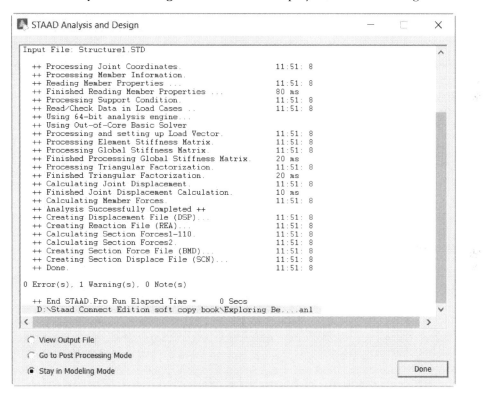

*Figure 7-7 The **STAAD Analysis and Design** window*

In this window, the STAAD.Pro will analyze the structural model and you can view the number of errors and warnings that were generated after the analysis. In this window, three radio buttons are available at the bottom. On selecting these radio buttons, you can view the results numerically, graphically, or you can stay in the modeling mode. The results include displacement, forces, stresses, support reactions, and so on. To remain in the modeling mode, select the **Stay in Modeling Mode** radio button. The other two radio buttons are discussed next.

View Output File

You can view the results numerically in the output file. To display the output file, select the **View Output File** radio button from the **STAAD Analysis and Design** window and then choose the **Done** button; the **STAAD Output Viewer** window will be displayed, as shown in Figure 7-8. This window will display the information related to the structure and their analysis results. When you scroll down in this window, you can view the post analysis results such as support reaction, member forces, element stresses, and so on. Figure 7-9 shows the support reactions displayed in the **STAAD Output Viewer** window.

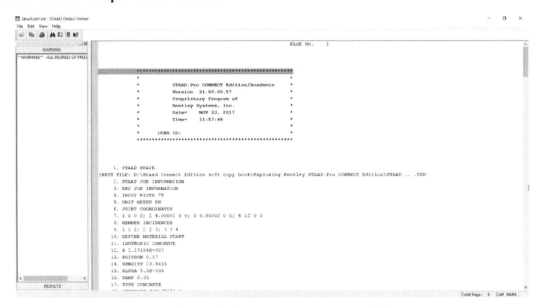

*Figure 7-8 The **STAAD Output Viewer** window*

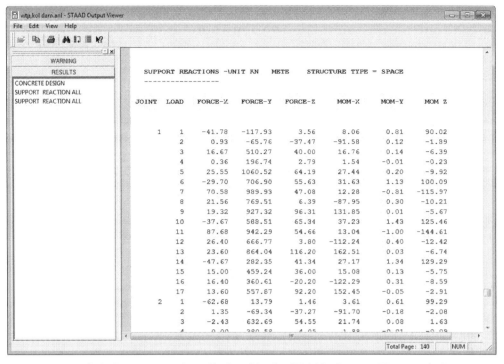

*Figure 7-9 Support reactions displayed in the **STAAD Output Viewer** window*

Go to Post Processing Mode

In the Post Processing mode, you can view the results graphically. These results include the bending moment diagrams, shear force diagrams, axial forces, deflections, and so on. To view the results graphically, select the **Go to Post Processing Mode** radio button in the **STAAD Analysis and Design** window and then choose the **Done** button; the **STAAD Analysis and Design** window will close and the **Results Setup** dialog box will be displayed, as shown in Figure 7-10.

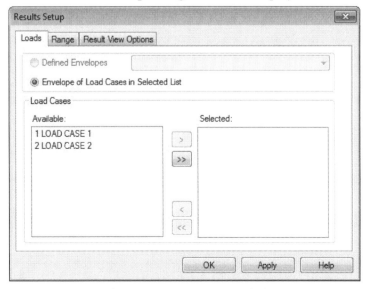

*Figure 7-10 The **Results Setup** dialog box*

In this dialog box, you can select the load cases and structural elements to be included in the post processing operations. In this dialog box, the **Loads** tab is chosen by default. In this tab, you will select the load cases to be included in the post processing operations. Select the required load cases from the **Available** area and then transfer them to the **Selected** area using the arrow buttons.

In the **Range** tab, you can specify the selection conditions based on the attributes for the members, nodes, properties, and so on. These selection conditions will determine whether the results will be displayed for all the members/nodes/properties or only for the selected ones.

In the **Result View Options** tab, you can specify the size for the displacement and moment diagrams to be plotted.

After selecting the required load cases and specifying other options, choose the **Apply** and **OK** buttons; the dialog box will be closed and the **Post Processing** tab will be displayed. In this tab, you can view the results graphically. These results are further categorized into further pages: **Displacements, Reactions, Beam Results, Plate Results, Solid Results, Dynamics, Reports.** Some of these tabs are discussed next.

Displacements

The **Displacements** page will be displayed by default in the **Post Processing** tab. In the **Displacement** page, the node deflection diagram for the selected load cases will be displayed in the main window. Figure 7-11 shows the node deflection diagram of a portal frame structure.

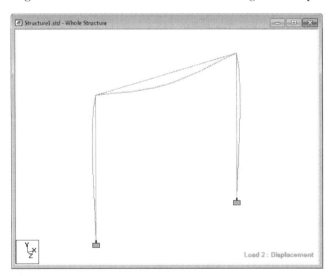

Figure 7-11 *Node deflection diagram in the main window*

The node displacement detail and the beam relative displacement detail of the selected load case will be displayed in the **Node Displacements** and **Beam Relative Displacement Detail** windows, respectively. These windows are available on the right side of the interface, refer to Figure 7-12.

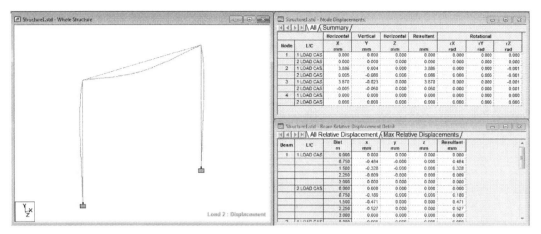

*Figure 7-12 The **Node Displacement** and **Beam Relative Displacement Detail** windows*

Reactions

You can also view the support reactions in the user interface screen. To do so, invoke the **Reactions** page; the support reactions will be displayed in a box at the nodes where support is provided, refer to Figure 7-13. If you are unable to view the support reactions properly, choose the **Select Text** tool from the **Selection** panel of the **Results** tab. Next, press and hold the left mouse button and drag the box displaying the reactions at the desired place in the user interface screen.

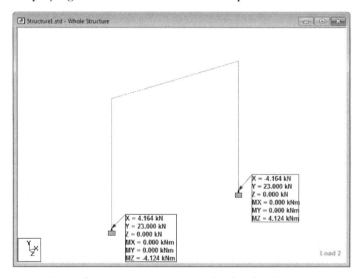

Figure 7-13 Support reactions displayed at the joints

Beam Results

You can view different types of member forces in the **Beam Results** page. On choosing this tab, the **Forces** page will be displayed. In this page, you can view the shear force diagram, bending moment diagram, and axial force diagram. The bending moment diagram of the structure is displayed by default, refer to Figure 7-14. You can toggle this bending moment diagram on or off by using the **Bending Moment about Z axis** button available in the **View Results** panel of the **Results** tab, as shown in Figure 7-15.

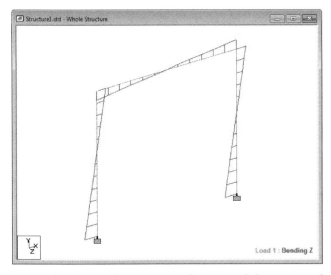

Figure 7-14 *Bending moment diagram of the structural members*

Figure 7-15 *Partial view of the* **View Results** *panel*

You can view the shear force and axial force diagrams by choosing the **Shear Y Force** and **Axial Force** buttons. Figures 7-16 and 7-17 show the shear force and axial force diagrams of a structure. The beam end forces results will be displayed in the **Beam End Forces** table on the right side of the interface. This table comprises three tabs: **All**, **Summary**, and **Envelope**. In the **All** tab, beam end forces of all the members at each end node will be displayed. In the **Summary** tab, maximum and minimum end force result for each degree of freedom will be displayed. In the **Envelope** tab, maximum positive and negative end force envelope along with the associated load case number will be displayed. In the **Beam Force Detail** table, maximum axial force, shear force, and bending moments will be displayed.

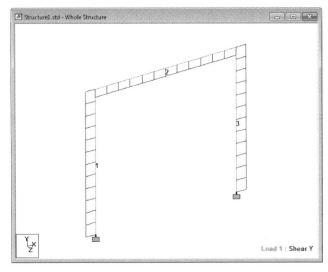

Figure 7-16 *Shear force diagram of the structural members*

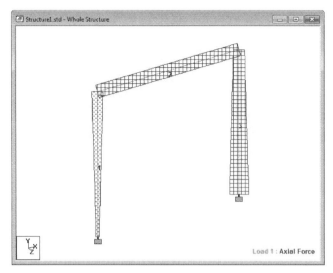

Figure 7-17 *Axial force diagram of the structural members*

From the **Beam Stress** page of the **Layouts** drop-down in the **View Results** panel of the **Results** tab; you can view the member stresses both graphically and numerically. The member stresses include combined axial and bending stresses. You can also view the combined stress for the cross-section at any point along the length of the member. When you invoke the **Stresses** page, the main window contains three sub windows: **3D Beam Stress Contour**, **Whole Structure**, and **Beam Combined and Axial Bending Stresses**. The **3D Beam Stress Contour** window is empty by default. In the **Whole Structure** window, the member stresses diagram will be displayed, as shown in Figure 7-18.

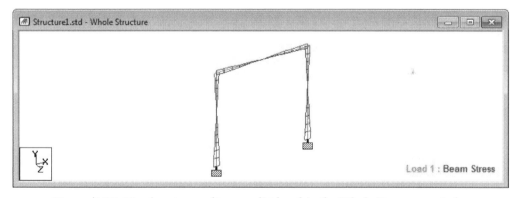

Figure 7-18 *Member stresses diagram displayed in the **Whole Structure** window*

The **Beam Combined Axial and Bending Stresses** window will comprises three tabs: **All**, **Max Stresses**, and **Profile Stress Points**. In the **All** tab, the combined axial and bending stresses at the four corners of the member cross section will be displayed. The maximum compressive and tensile stresses are also listed in this window. In the **Max Stresses** tab, maximum compressive and tensile stresses for all members of all load cases will be displayed. The **Profile Stress Points** tab is empty by default. In this tab, you can add forces and stresses (Combined Axial and Bending Stresses) obtained at certain selected points on a cross section. To do so, first select a member in the **Whole Structure** window. On doing so, the stress distribution of the selected member will be displayed in the **3D Beam Stress Contour** window and the **Select Section Plane** dialog box will also be displayed. The **3D Beam Stress Contour** window is divided into two panes. In the left pane, the combined stress distribution along the longitudinal axis of the member will be displayed along with a yellow section plane and in the right pane, stress distribution at the yellow section plane will be displayed, refer to Figure 7-19.

*Figure 7-19 Combined stress distribution in the **3D Beam Stress Contour** window*

In the **Select Section Plane** dialog box, you can use the slider to position the yellow rectangle at a certain location, refer to Figure 7-20. On doing so, the cross-sectional stress diagram will be immediately updated in the right pane of the **3D Beam Stress Contour** window.

Next, in the **Select Section Plane** dialog box, choose the **Add Stress to Table** button; the stresses will be added to the **Profile Stress Points** tab in the **Beam Combined Axial and Bending Stresses** window.

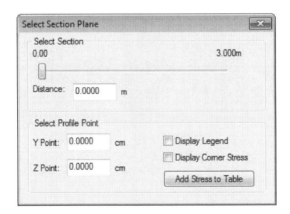

*Figure 7-20 The **Select Section Plane** dialog box*

You can also view the shear force, bending moment, and axial force diagrams of each member individually by choosing the **Graphs** option from the **Layouts** drop-down list of the **View Results** panel in the **Results** tab. On invoking the **Graphs** option, you will see the whole structure displayed in the main window and the member force diagrams displayed in the **Graph** area, refer to Figure 7-21. To view the graphs for a particular member, select it in the main window; the corresponding diagrams will be displayed in the **Graph** area.

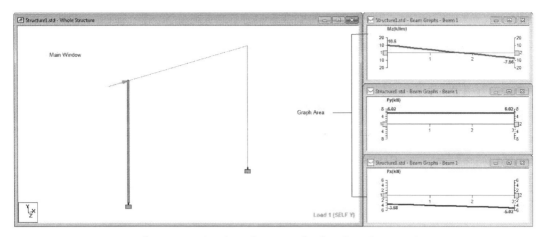

*Figure 7-21 Member force diagrams displayed in the **Graph** area*

You can view deflections, section displacements, mode shapes, and stresses in an animated mode by choosing the **Animation** button from the **Animation** panel of the **Results** tab. On doing so, the **Diagrams** dialog box will be displayed, as shown in Figure 7-22. In this dialog box, the **Animation** tab is chosen by default, refer to Figure 7-22. The options displayed in this dialog box are discussed next.

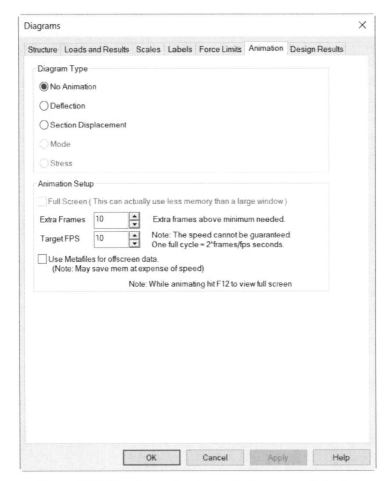

*Figure 7-22 The **Animation** tab in the **Diagrams** dialog box*

Select the type of animation by selecting the required radio button from the **Diagram Type** area. In the **Animation Setup** area, select the **Full Screen** check box in the **Animation Setup** area to display the animation in full screen rather than inside the window. To enhance the animation, specify the number of extra frames in the **Extra Frames** edit box. To speed up or slow down the animation, specify a value in the **Target FPS** edit box. To save the animated screens as windows metafiles, select the **Use Metafiles for offscreen data** check box.

 Note
*Selecting the **Use Metafiles for offscreen data** check box will make the speed of animation slower.*

PREPARING REPORTS

In STAAD.Pro, you can generate customized reports. These reports may include load cases, mode shapes, structural elements, numerical and graphical results, and so on. When you choose the **Reports** page of the **Post processing** tab; the **Report Setup** dialog box will be displayed, as shown in Figure 7-23. This dialog box comprises of tabs such as **Items**, **Load Cases**, **Modes**, **Ranges**, **Steel Design**, **Picture Album**, **Options**, **Name and Logo**, and **Load/Save**. These tabs are discussed next.

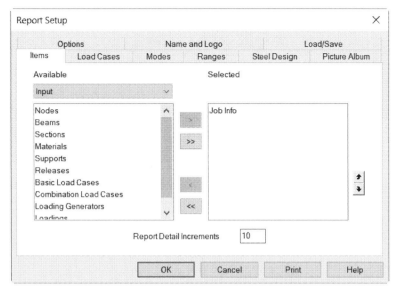

*Figure 7-23 The **Report Setup** dialog box*

Items

In the **Report Setup** dialog box, the **Items** tab is chosen by default. In this tab, you can specify the items to be included in the report. This tab is divided into two areas: **Available** and **Selected**. The **Available** area contains a drop-down list and a list box. Select an option such as **Input, Output**, or **Pictures** from the drop-down list available at the top; different items will be displayed in the list box. Next, select the items to be included in the report from the list box using the CTRL key and move them to the **Selected** area using the single forward arrow button. If you want to move the whole list from the **Available** area to the **Selected** area, choose the double forward arrow button.

Load Cases

In the **Load Cases** tab, you can specify which load case results will be included in the report. Then, you can move the required load cases from the **Available** area to the **Selected** area in the same way as discussed in the **Items** tab. You can also group the results table by node/beam numbers or by load cases. Specify the method of grouping using the radio buttons available under the **Grouping for Load Table**s and **Grouping for Result Tables** areas.

Modes

In the **Modes** tab, you can select the mode shape numbers for which the results will be displayed in the report. In this tab, select the required mode shape numbers from the **Available** area using the CTRL key and then move them to the **Selected** area.

Ranges

In the **Ranges** tab, you can specify the nodes, members, and elements to be included in the report. This can be done by selecting the following radio buttons available in this tab: **All**, **View, Group, Property**, and **Ranges**. These radio buttons are discussed next. You can select the **All** radio button to include all the members in the report. You can also generate a report for the structural element that will appear in a particular view. To do so, select the

View radio button and then select the required view from the drop-down list available next to it. If no view has been saved previously then the **View** radio button will be disabled. You can save a view by selecting the **View Management > Save View** from the **View** menu. The report for a particular group of members/elements can also be generated. To do so, select the **Group** radio button and then select the required group from the drop-down list available next to the **Group** radio button. To include members having certain property in the report, select the **Property** radio button. Next, select the required property from the drop-down list available next to the radio button. Similarly, you can generate a report for a particular set of nodes and members. To do so, select the **Ranges** radio button and then specify the range of nodes and members in the **Nodes** and **Beam/Plates/Solids** edit boxes, respectively.

Picture Album

You can also include the picture of the structure in the report. To do so, choose the **Picture Album** tab and then select the required picture from the **Name** drop-down list, refer to Figure 7-24. To print the picture on an entire page, select the **Full Page** check box or you can print the picture within the dimensions specified in the **Height** and **Width** edit boxes. To change the caption of the picture, specify a name in the **Caption** text box. To delete a picture, choose the **Delete Picture** button.

Note
*You can take a picture by using the **Take Picture** tool from the **Utilities** panel of the **Utilities** tab.*

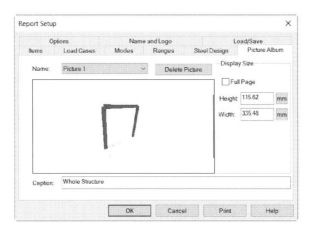

*Figure 7-24 The **Picture Album** tab in the **Report Setup** dialog box*

Options

In the **Options** tab, you can specify various report options. To include page headers, footers, and page outline in the report page, select the **Header**, **Footers, and Page Outline** check boxes, respectively. To include page numbers in the report, specify the starting page number in the **No. pages from** edit box. You can add prefix and suffix to the page numbers by specifying them in the **Prefix** and **Suffix** edit boxes, respectively. To reverse the page numbers, select the **Reverse page order** check box. Similarly, specify other options for the fonts and report table in the **Tables** area.

Name and Logo

In the **Name and Logo** tab, you can add your company's name and logo to the report. You can write the company's name in the viewing area, refer to Figure 7-25. You can change the font of the company name by using the options available in the **Text** area. To include company logo in the report, first save it in bitmap files (.bmp) format. Next, choose the **File** button in the **Graphics** area; the **Open** dialog box will be displayed. Browse to the file location, select the file, and then choose the **Open** button; the logo will be displayed in the **Viewing** area. You can place it at an appropriate place by selecting the **Left, Centre,** or **Right** radio button available in the **Graphic** area.

*Figure 7-25 The **Name and Logo** tab in the **Report Setup** dialog box*

Load/Save

In the **Load/Save** tab, you can save the contents of the report. You can also load the previously saved report using this tab. To save the report, choose the **Save As** button; the **Save Report** dialog box will be displayed. In this dialog box, specify the name of report in the **Save this report** text box and then choose the **OK** button; the report will be saved with the specified name and will be displayed in the **Report** drop-down list. To load a previously saved report, select the name of the report from the **Report** drop-down list and then choose the **OK** button. Choose the **Delete** button to delete the report.

After specifying all the parameters, choose the **OK** button; the report will be displayed in the main window. You can enlarge the viewing scale of the report. To do so, choose the **Zoom In** button available at the left side in the interface. To print the report, choose the **Print** button; the **Print** dialog box will be displayed. Specify the required settings in this dialog box to print the report or save it in Adobe PDF format. To make any changes in the report, choose the **Setup Report** button.

Note
In this chapter, you need to download the c06_Staad_connect.zip and c08_Staad_connect.zip files from http://www.cadcim.com. The path of the file is as follows: Textbook > Civil/GIS > STAAD. Pro > Exploring Bentley STAAD.Pro CONNECT Edition.

Example 1

In this example, you will open the file *c06_staad_connect_ex4.std* and then add the pre-print analysis, performing analysis, and the post analysis commands. Next, you will view the results and create a report file.

Steps required to complete this example are given below:

Step 1: Open the file *c06_staad_connect_ex4.std* in STAAD.Pro; the model is displayed in the main window.

Step 2: Choose **Analysis Commands** in the **Analysis Data** panel of the **Analysis and Design** tab; the **Analysis/Print Commands** dialog box is displayed. Select the **All** radio button from the **Perform Analysis** tab in the **Print Option** area and then choose the **Add** button. Choose the **Close** button to close the dialog box.

Step 3: Invoke **Post - Analysis Commands** in the **Analysis Data** panel of the **Analysis and Design** tab; the **Post Analysis Print - Whole Structure** window is displayed. Choose the **Define Commands** button available in the **Post Analysis Print - Whole Structure** window; the **Analysis/Print Commands** dialog box will be displayed, as shown in Figure 7-26.

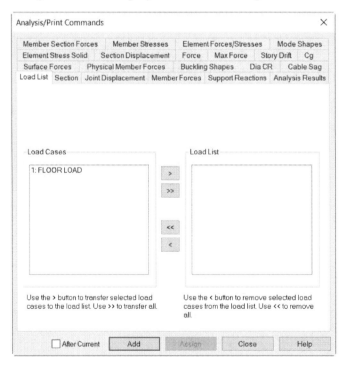

Figure 7-26 The Analysis/Print Commands dialog box

Step 4: In this dialog box, make sure that the **Load List** tab is chosen and then select the **1: FLOOR LOAD** in the **Load Cases** area. Next, choose the single forward arrow button; the **1: FLOOR LOAD** is moved to the **Load List** area. Choose the **Add** button; the **LOAD LIST** command will be added to the **Post Analysis Print - Whole Structure** window.

Step 5: Choose the **Member Forces** tab in the **Analysis/Print Commands** dialog box and select the **Global** check box. Then, choose the **Add** button to add the command to the **Post Analysis Print - Whole Structure** window.

Step 6: Choose the **Support Reactions** tab. Next, choose the **Add** button and then close the dialog box.

Step 7: Next, select the **PRINT MEMBER FORCES GLOBAL** command in the **Post Analysis Print - Whole Structure** window. Then, select the **Assign To View** radio button and choose the **Assign** button; the **STAAD.Pro CONNECT Edition** message box is displayed. Choose the **Yes** button; the command will be assigned to all the members.

Step 8: Next, select the **PRINT SUPPORT REACTION** command from the **Post Analysis Print - Whole Structure** window. Then, choose the **View Model From +Z** button from the **Tools** panel of the **View** tab; the front view of the structure is displayed, refer to Figure 7-27.

Step 9: Select **Nodes Cursor** from the **Selection** panel of the **Geometry** tab and then select the lower nodes, refer to Figure 7-27.

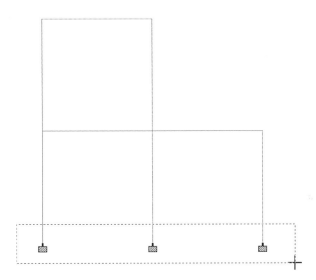

Figure 7-27 *Selecting the lower nodes using the drag box*

Step 10: Ensure that the **Assign To Selected Nodes** radio button is selected in the **Assignment Method** area of the **Post Analysis Print - Whole Structure** window. Choose the **Assign** button; the **STAAD.Pro CONNECT Edition** message box is displayed. Choose the **Yes** button; the command is assigned to the base nodes.

Step 11: Choose the **Isometric View** button from the **Tools** panel of the **View** tab; the isometric view of the structure is displayed in the user interface screen.

Step 12: Choose the **Run Analysis** option from the **Analysis** panel of the **Analysis and Design** tab; the **Warning** message box is displayed. In the **Warning** message box, choose the **Save** button; the **STAAD Analysis and Design** message box is displayed. Choose the **OK** button; the **STAAD Analysis and Design** window is displayed.

Step 13: In this window, select the **View Output File** radio button and choose the **Done** button; the **STAAD Output Viewer** window is displayed. In the left pane of this window, click on the **RESULTS** button and then click on **MEMBER FORCES GLOBAL ALL**; the member forces of all the members are displayed in the right pane of the window, refer to Figure 7-28.

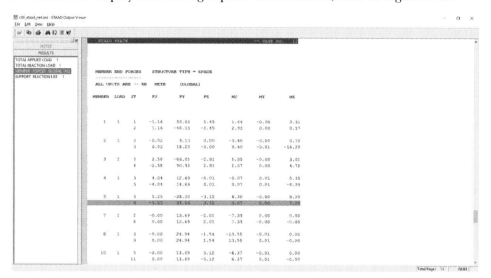

Figure 7-28 Member forces displayed in the STAAD Output Viewer window

Step 14: Repeat the procedure followed in step 13 and view the support reactions. Next, close the **STAAD Output Viewer** window.

Now, you will view the shear force and bending moment diagrams.

Step 15: Choose the **Post Processing** page from the left pane in the **Workflows** panel; the **Results Setup** dialog box is displayed. In this dialog box, choose the **Apply** and **OK** buttons; the **Displacements** page is displayed.

Step 16: Choose the **Beam Results** page from the **Postprocessing** tab; the **Forces** page is displayed with the bending moment diagram, as shown in Figure 7-29.

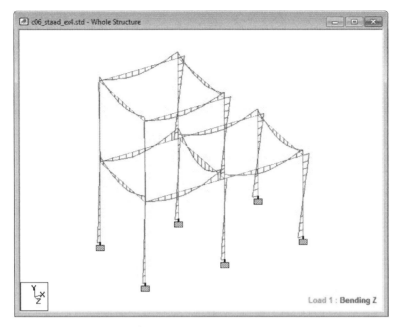

Figure 7-29 Bending moment diagram

Step 17: Choose the **Analytical Modelling** tab from the **Workflows** area and then choose the **Take Picture** tool from the **Utilities** panel of the **Utilities** tab. On doing so, the **Picture 1** dialog box is displayed and enter the text **BMD** in the **Caption** text box of this dialog box. Choose the **OK** button; the dialog box is closed and the picture is taken.

Step 18: Choose the **Bending Z Moment** button from the **View Results** panel of the **Results** tab; the bending moment diagram will disappear. Next, choose the **Shear Y Force** button from the **View Results** panel of the **Results** tab; the shear force diagram is displayed, as shown in Figure 7-30.

Step 19: Repeat the procedure followed in step 17 and take a picture with caption **SFD** of the shear force diagram.

Next, you will create a customized report for the structure.

Step 20: Choose the **Reports** page; the **Report Setup** dialog box is displayed. In the **Items** tab of the dialog box, select the **Input** option from the drop-down list available in the **Available** area. Next, in the **Available** area, select the **Nodes**, **Beams**, **Sections**, **Materials** and **Supports** option from the list box using the CTRL key and then choose the forward button; all the options are moved to the **Selected** area.

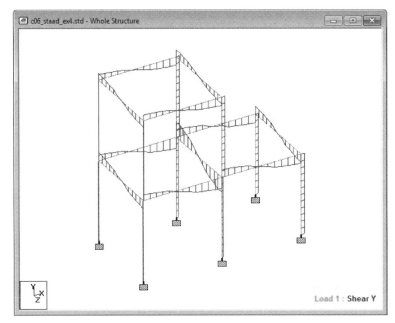

Figure 7-30 *Shear force diagram*

Step 21: In the **Available** area, select the **Output** option from the drop-down list. Next, select the **Beam End Forces** and **Reactions** options using the CTRL key in the list box and then choose the forward button; the options are moved to the **Selected** area.

Step 22: Similarly, select the **Pictures** option from the drop-down list available in the **Available** area and move **Picture 1** and **Picture 2** to the **Selected** area.

Step 23: Choose the **Load Cases** tab and ensure that the **1 FLOOR LOAD** is available in the **Selected** area.

Step 24: Choose the **Name and Logo** tab and enter **CADCIM TECHNOLOGIES** in the Viewing area. Next, choose the **File** button in the **Graphics** area; the **Open** dialog box is displayed.

Step 25: Browse to the folder *c07_Staad_connect* and then select the *logo.bmp* file. Next, choose the **Open** button; the logo is added to the Viewing area. Ensure that the **Left** radio button is selected in the **Text** area and the **Right** radio button is selected in the **Graphic** area. Figure 7-31 shows the name and logo added to the Viewing area.

Step 26: Choose the **Load/Save** tab and then choose the **Save As** button; the **Save Report** dialog box is displayed. Specify the name *Ex1_report* in the **Save this report as** text box and then choose the **OK** button; the report name is added to the **Report** drop-down list.

Step 27: Choose the **OK** button in the **Report Setup** dialog box. Now, choose the **Print** button from the right side of the interface; the **Print** dialog box is displayed. In this dialog box, select the **Adobe PDF** option from the **Name** drop-down list and then choose the **OK** button; the **Save PDF File As** dialog box is displayed. Browse to an appropriate location and save the file with the name *c07_Staad_connect_ex1* and then choose the **Save** button; the report is saved as a pdf file, refer to Figure 7-32.

Figure 7-31 *Figure and logo added in the Viewing area*

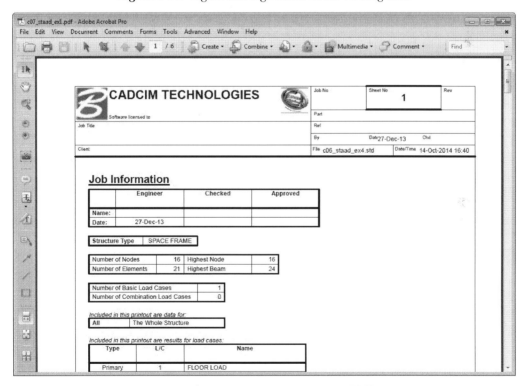

Figure 7-32 *Report file saved as a pdf file*

Step 28: Choose the **Save As** option from the **File** menu; the **Save As** dialog box is displayed. In this dialog box, specify the name *c07_staad_connect_ex1* in the **File name** edit box and save it at an appropriate location.

Step 29: Close the file by using the **Close** option in the **File** menu.

Self-Evaluation Test

Answer the following questions and compare them to those given at the end of this chapter:

1. The _____ command is used to print basic information about the structure in the output file.

2. The _____ command is used to print entire information about the structure in the output file.

3. You can use the _____ analysis tab for the structures subjected to lateral loading.

4. You cannot print mode shape values in the output file. (T/F)

5. In the post processing mode, you cannot view the results graphically. (T/F)

6. You can view the shear force diagram, bending moment diagram, and axial force diagram of each member individually. (T/F)

7. You can toggle on/off the shear force diagram and bending moment diagram of whole structure. (T/F)

8. You can graphically view the stress contours of members and plates. (T/F)

9. You cannot insert pictures of the structural model in the report. (T/F)

10. You can add the logo of a company with the file extension other than *.bmp*. (T/F)

Review Questions

Answer the following questions:

1. Which of the following tabs is used to print the coordinates of the centre of gravity of the structure in the output file?

    (a) **Force**   (b) **Section Displacement**
    (c) **Cg**   (d) None of these

2. Which of the following tabs is used to calculate the section displacement and print it in the output file?

    (a) **Section Displacement**   (b) **Force**
    (c) **Member Stresses**   (d) **Mode Shapes**

3. Which of the following tabs is used to calculate and print the support reactions in the output file?

    (a) **Member Forces**   (b) **Support Reactions**
    (c) **Story Drift**   (d) None of these

4. You cannot add the logo of a company to the STAAD report. (T/F)

5. In STAAD.Pro, you cannot create animation of deflection and mode shapes. (T/F)

Answers to Self-Evaluation Test

1. PRINT PROBLEM STATISTICS, 2. PRINT ALL, 3. PDelta, 4. F, **5.** F, **6.** T, **7.** T, **8.** T, **9.** F, **10.** F

Chapter 8

Physical Modeling

Learning Objectives

After completing this chapter, you will be able to:
• *Create grids, nodes, members using Physical Modeller*
• *Create supports using Physical Modeller*
• *Create load details using Physical Modeller*
• *Generate an analytical model from physical model*

INTRODUCTION

In analytical modeling, the formation of structure starts with smallest unit of elements such as nodes and then proceeds with members, surfaces, plates, assigning materials and sections to them.

In STAAD.Pro CONNECT Edition, physical modeling starts with modeling the structures using physical concepts. It has an advantage of simplifying the modeling of a structure. It reflects the process of building a model accurately.

STARTING A NEW PROJECT IN PHYSICAL MODELING

To start a new project, choose the **New** option from the left pane; the **New** page with the **Model Information** page will be displayed. Specify the name and location of the file in the **File Name** and **Location** edit boxes, respectively. Next, choose **Physical** in the **Type** area and **Metric** in the **Units** area. After specifying all the data, choose the **Create** option in the **New** page; a new project will start and the user interface will be displayed, as shown in Figure 8-1.

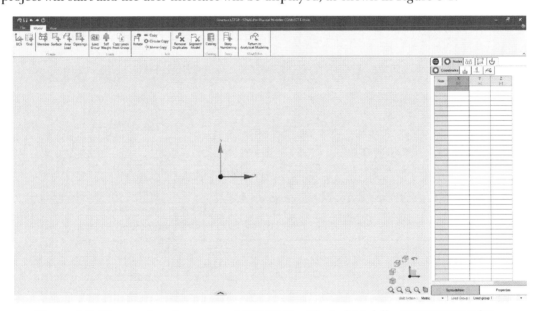

Figure 8-1 *The user interface screen of STAAD.Pro Physical Modeller CONNECT Edition*

Now you need to specify the unit system and set the units for different properties such as length, force, and so on. To do so, choose the **Options** option from the **File** menu; the **Options** dialog box will be displayed with the **Units** tab chosen by default, refer to Figure 8-2. In this tab, the **Units configurations** area displays different properties and their respective units based on the default unit system selected in the **Units system** drop-down list. Next, select the desired unit system from the **Units system** drop-down list. On specifying the unit system, the default units of the properties in the **Units configuration** area will also change accordingly. Choose the **OK** button; the dialog box will be closed and unit system will be changed in the Status Bar. You can also change the unit system and load group for the current model from the Status Bar by selecting the required option from the corresponding drop-down lists.

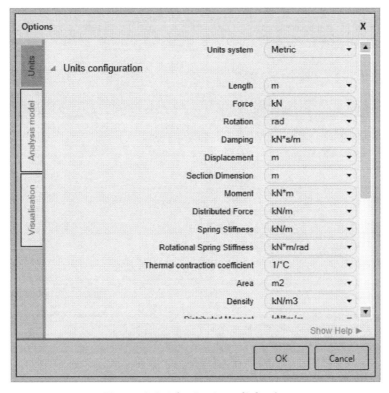

*Figure 8-2 The **Options** dialog box*

MODELING

In STAAD Pro, for modeling a structure, you first need to select the coordinate system, then draw the structure, assign model properties, define boundary conditions, and finally apply loads to the structure. The physical model is then transferred to analytical modeling for design analysis. The processes involved in physical modeling of structures are discussed in detail next.

Using Construction Aids

Using construction aids, you can generate a coordinate system such as **World Coordinate System (WCS)** or **Three points**, as per your requirement. Additionally, you can create construction grid in any plane with the required spacing and the number of construction lines along any axis.

Adding Coordinate System

STAAD.Pro allows you to use multiple coordinate systems or the Unified Coordinate System (UCS), within the same model. An initial UCS is created by default in each model. By default, this UCS is set at the global origin with no rotation. The coordinate systems can be translated or rotated with respect to the WCS. To add a coordinate system, choose the **UCS** tool from the **Create** panel of the **Modify** tab; the **Create UCS** dialog box will be displayed, as shown in Figure 8-3.

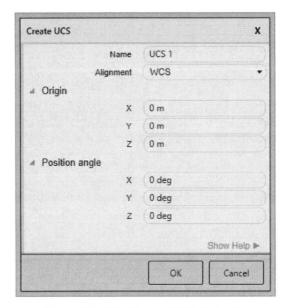

Figure 8-3 The Create UCS dialog box

In this dialog box, select the required option from the **Alignment** drop-down list; the options will be updated accordingly. Enter the coordinate values in the **Origin** area and the angle values in the **Position angle** area. Now, on selecting the **Three points** option from the **Alignment** drop-down list; the options in the **Create UCS** dialog box will change accordingly. Enter the values of the coordinates in the **Origin** area and the coordinates of X and Y axes in the **Position X axis** area and the **Position Y axis** area. Choose the **OK** button; the dialog box will be closed and the UCS will be added. To toggle on or off the display of coordinates system, choose the **UCS** tool from the **View** tab.

Adding Construction Grid

To add a grid and display it in all view windows, choose the **Grid** tool from the **Create** panel of the **Model** tab; the **Create Grid** dialog box will be displayed, as shown in Figure 8-4. Select the required UCS from the **Reference CS** drop-down list and then select the **UCS** to place the grid origin. Select the required option from the **Creation method** drop-down list for the type of spacing and dimension in the X and Y axis, respectively. For specifying spacing, use the format number@distance. You can specify different spacings by separating the values using comma. Also, you can set the units by using their abbreviations. Different units of length can be used. Otherwise, the default unit of length is assumed. For example, 3@10 cm, 2@5 ft can be typed for three spaces at ten centimeters, followed by two spaces at five feet. Choose the **OK** button; a new grid will be added.

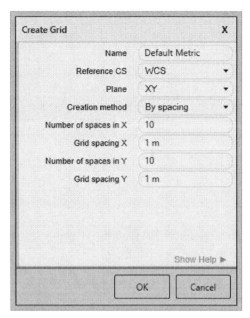

*Figure 8-4 The **Create Grid** dialog box*

Drawing the Structure

In STAAD.Pro Physical Modeller, you can also draw members, surfaces and nodes. This section will explain how to use graphical drawing tools to draw members, surfaces and nodes.

Drawing the Members

This section explains how to draw the members in a construction grid or draw the members between the existing nodes. These are discussed next.

Drawing a Member in a Construction Grid

To draw a member in a construction grid choose the **Member** tool from the **Create** panel of the **Model** tab; the tool is highlighted indicating that it is active. Also the mouse pointer changes to an add member cursor and a yellow symbol "snaps" to the nearest snap point. Click at a point to start a node of the new member; a dashed line is attached from this point to the mouse pointer indicating where the member will be placed with the next mouse click. Click the end node for the new member; the member is drawn in place as a line model, refer to Figure 8-5. Again repeat the same procedure to add as many more connected members as needed. The previous member's end node is automatically selected as the next member's start node. This allows you to keep drawing continuous members. Double-click on the final node of the member to end the line.

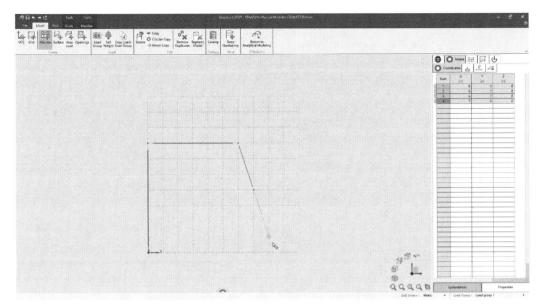

Figure 8-5 *Creating members in the grid*

Note
*You can double-click at the end node to exit the current selection in a series of members. To deactivate the tool either select a different tool or the **Member** tool again, or press the ESC key.*

Generating Members Between Existing Nodes

To create members from the existing node, select two or more nodes and then choose the required direction from the **Member Generation** drop-down of the **Edit** panel in the **Node** tab; the members will be created. Note that the members are generated along the selected direction from the first to last nodes in the selected direction.

Drawing Surfaces

Now you need to add a surface to the model and then add opening in the existing surface. These processes are discussed next.

Adding a Physical Surface

To add planar structural object to the physical model, add a grid and then choose the **Surface** tool from the **Create** panel of the **Model** tab; the tool is highlighted indicating that it is active. Also the mouse pointer changes to an add surface cursor, and a yellow symbol snaps to the nearest snap point. Click on the point to define the start point for the new surface and then click as many times as the corners are needed. Continue clicking clockwise or counter-clockwise to define the surface such that no edge of the surfaces crosses any previously defined edge, refer to Figure 8-6.

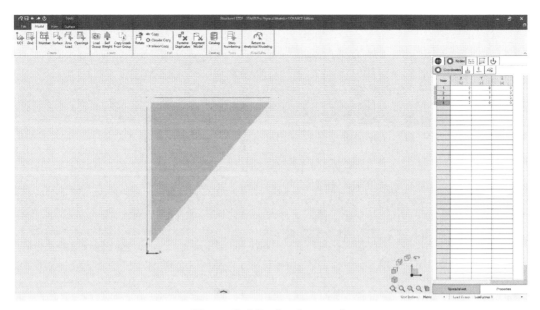

Figure 8-6 Surface is created

Adding an Opening in a Surface

To draw an opening in an existing surface, choose the **Openings** tool from the **Create** panel of the **Model** tab; the tool is highlighted indicating that it is active and the mouse pointer changes to an add openings cursor. Click at a point on surface to define the start point for the new opening and then click the subsequent corners to define the opening, continuing in clockwise or counter-clockwise order. Click as many corners as needed, so long as no edge of the opening crosses any previously defined opening edge, refer to Figure 8-7.

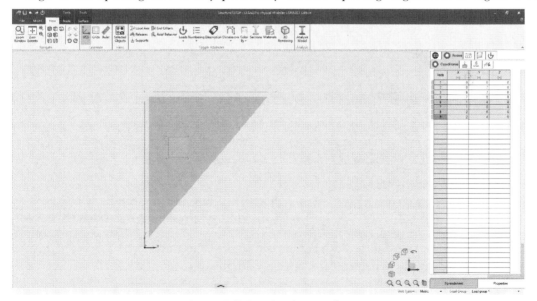

Figure 8-7 Opening in a surface

Adding Nodes

STAAD.Pro Physical Modeller contains features to generate additional nodes in addition to manually adding nodes. For many models, you will not need to directly add nodes. They are automatically generated in the physical model at the ends of drawn members and at the vertices of surfaces and surface openings.

Adding Nodes Manually

Physical nodes are typically added by drawing members or surfaces or making edits to the object types. You can manually add nodes by choosing the **Coordinates** option from the **Nodes** tab in the worksheet area and then entering the global coordinates for the new node in the first blank row of the X, Y, and Z columns, respectively. The current length units are displayed in the column headings, as shown in Figure 8-8. You can manually type different units. Such units are automatically converted into the current units once you have finished entering the value. You can copy and paste multiple values from external spreadsheet programs into the worksheet. You can press the **Tab** button to move to the next cell in the row or press ENTER to move to the next row. A new node number is automatically assigned to the row when you begin typing coordinate values.

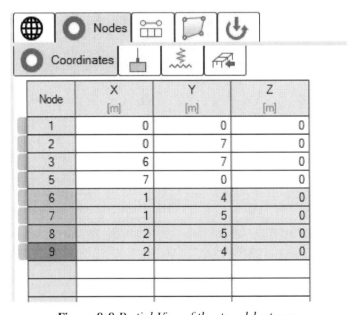

Figure 8-8 Partial View of the spreadsheet area

Generating Nodes Along a Line

To add a line of evenly spaced nodes, first select any two nodes and then choose the **Linear Generation** tool from the **Edit** panel of the **Node** tab; the **Linear Generation of Nodes** dialog box will be displayed, refer to Figure 8-9. You can enter the number of nodes to add between the two selected nodes in the **Nodes to generate between selection** edit box. Choose the **OK** button; the dialog box will be closed and the specified number of nodes will be added between the two selected nodes.

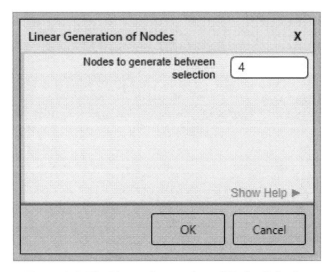

*Figure 8-9 The **Linear Generation of Nodes** dialog box*

Generating Nodes in an Arc

To add a circular arc of evenly spaced nodes, you need to select the group of three nodes in the model. The first node you select will be the center of the circle and the second node selected creates a vector with the first node perpendicular to the circle. Hence, the third node is the starting point of the arc and thus determines the radius of the arc. The new nodes will lie in the plane of nodes 1-3 and perpendicular to nodes 1-2. In order to form two planes, the three selected nodes cannot lie on a single line. To create the circular arc of nodes, select the **Circular Generation** tool from the **Edit** panel of the **Node** tab; the **Circular Generation of Nodes** dialog box will be displayed, as shown in Figure 8-10. Specify the number of nodes and the arc angle in the **Number of Nodes** and **Arch angle** edit boxes, respectively. Choose the **OK** button; the dialog box will be closed and the nodes will be generated along the specified arc.

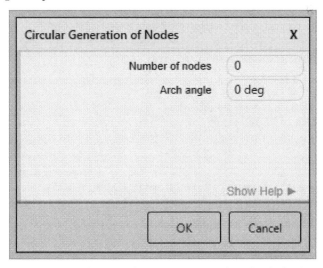

*Figure 8-10 The **Circular Generation of Nodes** dialog box*

Generating Nodes in a Rectangular Array

To add a quadrangular array of nodes, choose a group of four nodes in the model. Select nodes within single plane (not along single straight line) to define the plane of node generation. The line segments between nodes do not need to be parallel. They can form any four-sided polygonal shape. Choose the **Quadratic Generation** tool from the **Edit** panel of the **Node** tab; the **Quandrangular Generation of Nodes** dialog box will be displayed, as shown in Figure 8-11. Enter the number of nodes along the perpendicular sides. The sides will be labeled by the order of the selection of nodes. Next, choose the **OK** button; the dialog box will be closed and the nodes will be generated in the selected plane.

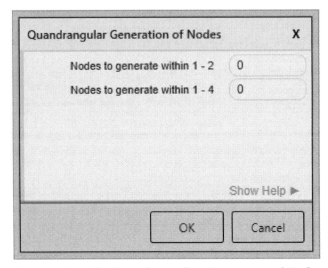

Figure 8-11 The *Quandrangular Generation of Nodes* dialog box

Example 1

Create a 2D portal frame with coordinates (0,0,0), (0,15,0), (20,15,0), (20,0,0).

Steps required to complete this example are given below:

Step 1: Start STAAD.Pro CONNECT Edition, and select the **New** option from the left pane of the user interface screen; the **New** page is displayed with the **Model Information** page. Enter *c08_staad_connect_ex1* in the **File Name** edit box and browse to the location *C:\STAAD Examples\ c08_staad_connect* by clicking the button next to the **Location** edit box.

Step 2: Select **Physical** from the **Type** area and **Metric** from the **Units** area and then choose the **Create** button from the **Model Information** tab; the user interface screen is displayed in the STAAD.Pro Physical Modeller. Note that the units can be changed later if necessary, at any stage of the model creation.

Step 3: Choose the **Grid** tool from the **Create** panel of the **Model** tab; the **Create Grid** dialog box will be displayed. Enter **Grid 1** in the **Name** edit box and keep default settings same. Enter **20** in the **Number of spaces in X** edit box and **1** in the **Grid spacing X** edit box. Next, enter

15 in the **Number of spaces in Y** edit box and **1** in the **Grid spacing Y** edit box, as shown in Figure 8-12. Choose the **OK** button; the dialog box is closed and the grid is created in the drawing area, as shown in Figure 8-13.

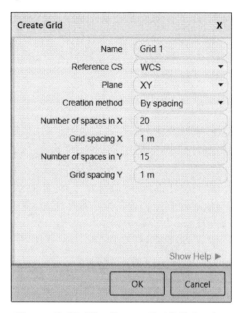

*Figure 8-12 The **Create Grid** dialog box*

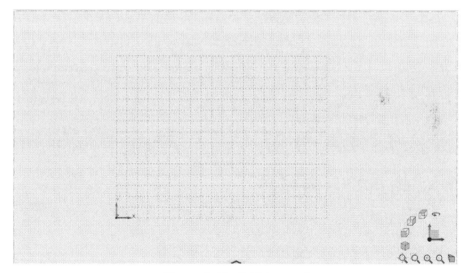

Figure 8-13 Grid 1 created

Step 4: Choose the **Member** tool from the **Create** panel of the **Model** tab; the tool is highlighted with a yellow circle attached to the cursor and snaps to the grid intersection when you move it to the view area.

Step 5: Select the lower left corner of the grid (the origin at 0,0,0); a dashed line is attached to the cursor. Select the following two points around the edge to draw two members: (0,15,0), (20,15,0). Then, double-click on the final point at (20,0,0). This will stop the dashed line at that point and you have connected all the members. Choose the **Member** tool again to make it inactive.

Step 6: Again, choose the **Grids** tool from the **Reference** panel of the **View** tab; the grid is no longer highlighted and all grid lines are hidden in the display. Next, choose the **Node** and the **Member** options from the **Numbering** drop-down in the **Model** panel of the **View** tab to display the node and beam labels in the drawing area. Next, choose the **Numbering** tool and select the entire structure from the user interface screen to highlight it, as shown in the Figure 8-14.

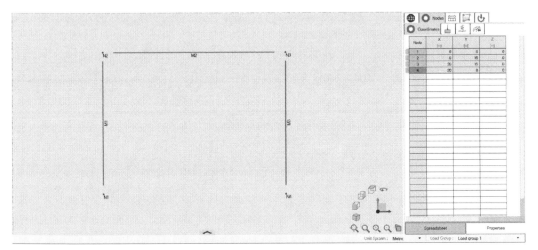

Figure 8-14 2D Portal Frame

Step 7: Choose the **Create analysis model** option from the **File** menu; the **Return to Analytical Modeling** message box is displayed with the warnings. Choose the **OK** button; the **Create analysis model** message box is displayed with the message that the process was completed with warnings. Choose the **Yes** button; the Physical Modeller is closed and the model is generated and opened into analytical modeler. Choose the **Save As** option from the **File** menu and save the file to the desired location. Choose the **Close** option from the **File** menu; the file is closed.

Assigning Model Properties

After you have drawn a model, you must assign material, section and thickness to members, and surfaces. This section explains how these properties are assigned to the members, surfaces.

Assigning Material to Members

To assign a material to the member, select the member and choose the **Material** tool from the **Assign properties** panel of the **Member** tab; the **Assign material** dialog box will be displayed, as shown in Figure 8-15. Choose the required source to assign the member from the **Source** drop-down list. The **Catalog** option is used to select materials from the Section Property Catalog, the **Model** option is used to select the materials previously assigned to your current model, and the **User Database** option is used to select custom materials previously saved in your user database. Then, choose the type, country, specification, and name as required from the dialog box.

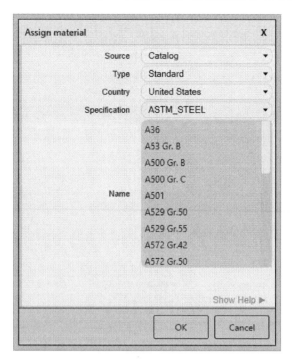

Figure 8-15 *The **Assign material** dialog box*

Next, choose the **OK** button; the dialog box will be closed and the material selected will be assigned to the selected member.

Assign Section to Members

After assigning material to the model, you need to provide a section to the member. You can assign a previously added section or a new section to your model and also a prismatic section to your model which are discussed next.

Assigning a Section to Members

To assign a section to a member, select the member and then choose the **Section** tool from the **Assign properties** panel of the **Member** tab; the **Assign section** dialog box will be displayed, as shown in Figure 8-16. You can assign a previously defined section or add a new section for use. Choose the section you want to assign by selecting the appropriate material, category, country, specification, version, manufacturer, table, and name from their corresponding drop-down list. Next, choose the **OK** button; the dialog box will be closed and the selected section will be assigned to the selected member.

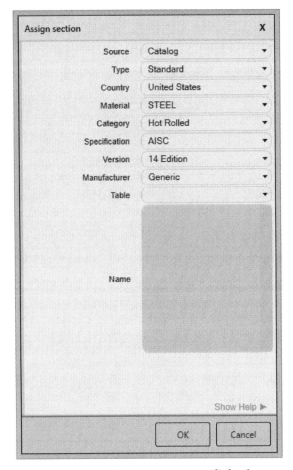

Figure 8-16 The **Assign section** *dialog box*

Assigning a Prismatic Section

To add a prismatic section to the model and assign it to the current selected members, select the member which you want to assign the section and then choose the **Section** tool from the **Assign properties** panel of the **Model** tab. On doing so, the **Assign section** dialog box will be displayed, refer to Figure 8-16. Choose the **Prismatic** option from the **Type** drop-down list; the options will change accordingly. Select the required option from the **Shape** and **Template** drop-down lists, respectively. Enter the dimensions for the selected template in the corresponding cell for each dimension. Type a unique section name in the **Name** edit box to identify the section in this model. Choose the **OK** button; the dialog box will be closed and the selected section will be assigned to the selected member.

Assigning Material to a Surface

To assign a material to a surface, select the surface on which you want to assign the material. Choose the **Material** tool from the **Assign Properties** panel of the **Surface** tab; the **Assign material** dialog box will be displayed, as shown in Figure 8-15. In this dialog box, select the required option from the **Source** drop-down list for the material definition and accordingly the options displayed in the dialog box will vary. Enter the required values in their corresponding edit boxes and then choose the **OK** button; the dialog box will be closed and the material will be assigned to the selected surface.

Assigning Thickness to a Surface

To assign thickness value to surface, select the surface and then choose the **Thickness** tool from the **Assign properties** panel of the **Surface** tab; the **Assign Surface Thickness** dialog will be displayed, as shown in Figure 8-17. Enter the value of the thickness in the **Thickness** edit box and then choose the **OK** button; the dialog box will be closed and the thickness will be assigned to the selected surface.

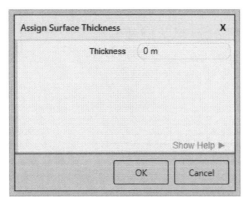

Figure 8-17 *The Assign Surface Thickness dialog box*

Example 2

In this example, you will create a portal frame with coordinates as (0,0,0), (0.15,0), (20,15,0), (20,0,0) and name it as c08_staad_connect_ex2.std file. Next, you will assign member properties to the portal frame.

Steps required to complete this example are given below:

Step 1: Create the portal frame as discussed in Example 1 of Chapter 8. Save the file by choosing the **Save As** option from the **File** menu and enter *c08_staad_connect_ex2* in the **File Name** edit box. Next, browse to the location *C:\STAAD Examples\c08_staad_connect* by clicking the button next to the **Location** edit box and choose the **Save** button.

Step 2: Select the two column members (M1 and M3) and click anywhere in the empty area of the view window; all model objects (members and nodes) are deselected. Next, click and drag the cursor horizontally from right to left across the columns to select them.

Step 3: Choose the **Section** tool from the **Assign properties** panel of the **Member** tab. The **Assign Section** dialog box will be displayed. Specify the section data for the columns, as shown in Figure 8-18. Next, choose the **OK** button; the section is assigned to the columns.

Figure 8-18 *The **Assign section** dialog box*

Step 4: Select the beam and ensure that all other members are deselected. Repeat steps 1 and 2 and assign a section **ISMB 200** to the beam (M2).

Step 5: Now, select all three members and choose the **Material** tool from the **Assign properties** panel of the **Member** tab; the **Assign material** dialog box is displayed. Specify the material data for the members, refer to Figure 8-19.

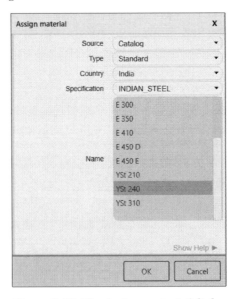

Figure 8-19 *The **Assign material** dialog box*

Step 6: Choose the **OK** button; the material is assigned to the members, as shown in Figure 8-20. Choose the **Save** option in the **File** menu to save the file. Then, close the file.

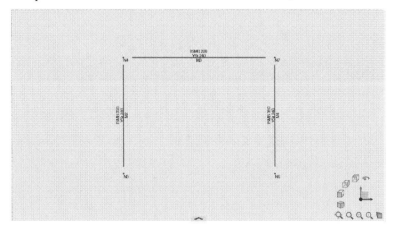

Figure 8-20 Portal frame with assigned sections and material

Defining Boundary Conditions

Boundary conditions are necessary for analysing and designing a structure. Supports could be fixed, pinned, or customized. This section explains how to assign the support conditions in physical modeling. Following types of supports can be assigned to the model in physical modeling.

Assigning Fixed or Pinned Support

STAAD.Pro Physical Modeller includes predefined boundary conditions for fixed and pinned nodal supports. Fixed supports are restrained in all six degrees of freedom (TX, TY, TZ, RX, RY, and RZ). Pinned supports are restrained against translation (TX, TY, and TZ) but free to rotate (i.e., released in RX, RY, and RZ). To assign fixed or pinned support, select the nodes to which you want to assign supports. Choose one of the following tools according to the support condition you want to use: the **Fixed** tool or the **Pinned** tool from the **Supports** panel of the **Node** tab; the supports would be assigned.

Assigning Custom Nodal Support

To assign a custom nodal restrain or spring, select the nodes to which you want to assign the same support type. Next, choose the **Custom** tool from the **Supports** panel of the **Node** tab; the **Assign Custom Support** dialog box will be displayed, refer to Figure 8-21. Select the **Restraints** or **Springs** support type. For **Restraints**, check each degree of freedom which should be fixed against **Translation** or **Rotation**. The dialog box will be closed and update the supports based on the selection. For **Springs**, select the **Axial behavior** of the spring. Select the global axis along which the axial behavior acts (applies to tension-only and compression-only springs). Enter a **Translation** and **Rotation** spring value for each direction in which the spring acts. Next, choose the **OK** button; the custom support is assigned to the selected support.

Assigning Supports to Members

To assign supports at evenly spaced points along a member, select the members which have the same number of supports along their length. Choose the **Custom** tool from the **Supports** panel of the **Member** tab; the **Assign Custom Linear Support** dialog box will be displayed, refer to Figure 8-22. Select the support type from the **Support Type** drop-down list. Next, enter the number of segments to be used along the length of the members from the **Segments** edit box. For **Restraints**, check each degree of freedom which should be fixed against **Translation** or **Rotation**. For **Springs**, choose the **Axial behavior** of the spring. Choose the global axis along which the axial behavior acts (applies to tension-only and compression-only springs). Next, enter **Translation** and **Rotation** spring values for each direction in which the spring acts. Choose the **OK** button; the dialog box will be closed and the support will be assigned based on the selection.

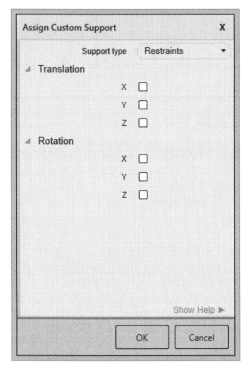

Figure 8-21 The Assign Custom Support dialog box

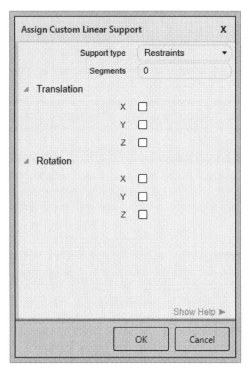

Figure 8-22 The Assign Custom Linear Support dialog box

Assigning Mat Foundation

To assign a mat foundation support to a surface, select one or more surfaces that will be supported. Choose the **Mat Foundation** tool from the **Mat and Edge Supports** panel of the **Surface** tab; the **Assign Mat Foundation** dialog box will be displayed, refer to Figure 8-23. Specify the direction in the **Direction** edit box in which the foundation reaction acts. Enter the Subgrade modulus value in the **Subgrade modulus** edit box for the soil modeling. Select the required option from the **Axial Behavior** drop-down list for the capability of mat foundation to resist the force along the selected direction. Next, choose the **OK** button; the dialog box will be closed and the mat foundation will be assigned to the selected surface.

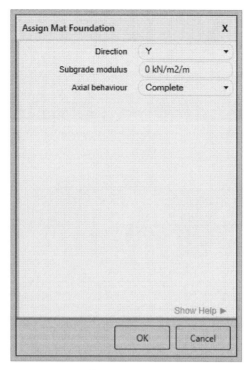

Figure 8-23 The **Assign Mat Foundation** *dialog box*

Applying Loads

In the previous sections, you learnt to draw a model, assign properties to its members as well as surfaces and plates, and define boundary conditions. Now, you will learn to apply loads to the model. The following section describes methods to work with load groups and assign different loads to the physical model.

Working with Load Groups

You can group the loads in STAAD.Pro Physical Modeller using the Load Groups. Each load item created will be assigned to the current load group, which is set in the program window status bar. When the physical model is moved to the analysis model, each load group generates a reference load case in the STAAD input file. These load cases can then be used within primary load cases in any combination needed. The reference load cases are not directly analyzed. This section describes how to create, edit, and use load groups in your physical model.

Creating a Load Group

To add a new load group to your model, choose the **Load Group** tool from the **Loads** panel of the **Model** tab; the **Add Load Group** dialog box will be displayed, refer to Figure 8-24. Enter a name for the new load group in the **Name** edit box and type a description for the load group in the **Description** edit box. Next, select the load type from the **Type** drop-down list. Enter **Self-weight** multipliers in each global direction if the load case contains self weight. Then, choose **OK** button; the load group will be created and selected as the current load group.

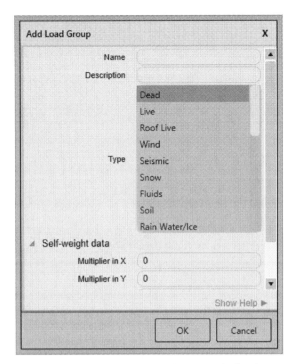

Figure 8-24   *The **Add Load Group** dialog box*

Adding Self Weight to a Load Group

Staad.Pro Physical Modeller considers self-weight of the members, plates, and solids. The self-weight of every element will be calculated and applied as a uniformly distributed member load. The self weight of plates are placed at the plate joints. To add self weight of an element to the current load group, choose the **Self Weight** tool from the **Loads** panel of the **Model** tab; the self weight multiplier of value -1 will be applied in the Y direction to the current load group.

Copying Loads from a Load Group

To copy the load items from another load group to the current load group, choose the **Copy Loads From Group** option from the **Loads** panel of the **Model** tab; the **Copy Load Group** dialog box will be displayed, refer to Figure 8-25. Enter the name of the load group in the **Copy loads from** edit box. Choose the **OK** button; all load items from the source load group will be copied to the current load group (selected in the status bar).

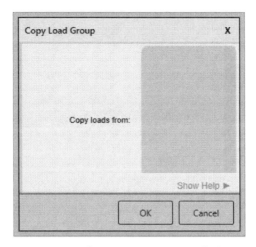

Figure 8-25 *The **Copy Load Group** dialog box*

Nodal Loads

Nodal loads are the loads that can be applied to the nodes. Nodal load consists of concentrated loads. Thes loads can be applied in in the X, Y, or Z direction. This section explains how these loads are assigned to the physical model.

Apply Nodal Load

Load items are added to the current load group. Nodal loads (i.e. joint loads) – both forces and moments – may be applied to any free joint of a structure. You can apply multiple loads on a single node. To assign a concentrated load or moment at a node, select the nodes which have the same load. Choose the **Nodal Load** tool from the **Loads** panel of the **Node** tab; the **Add Nodal Load** dialog box will be displayed, as shown in Figure 8-26. Choose the load group from the **Load group** edit box. Enter the force and moment magnitudes acting along the global directions at the nodes in the **Fx**, **Fy**, **Fz**, **Mx** edit boxes. Positive force values are taken in the direction of the global axes. Positive moment is taken in counter-clockwise direction about the global axes. Next, choose the **OK** button; the dialog box will be closed and the loads will be assigned to the nodes.

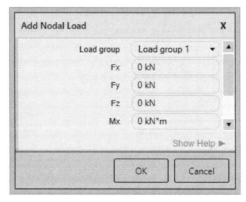

Figure 8-26 *The **Add Nodal Load** dialog box*

Member Loads

Different types of member loads can be assigned to a model. Some of these loads are discussed next.

Applying Concentrated Load

To assign a concentrated load or moment to the selected member, choose the required load from the **Loads** panel of the **Member** tab; the dialog box corresponding to the type of load assigned will be displayed. Now, select the direction in which the loads will be applied. In case of concentrated load, the force will act perpendicular to the direction of the member. For concentrated moment, the moment will act along the axis of the member. Select the **As a percentage** check box to input distances as a percentage of the overall member length. Enter the load magnitude and location values in their corresponding edit boxes. Select the **Loads** drop-down list. A table will be displayed to enter the values.

For each concentrated load or moment, enter the magnitude and distance. Negative values may be entered for magnitude of Load or Moment. For concentrated forces, enter a transverse distance value to apply the load through a point other than the shear center. This eccentric loading will result in a moment in addition to force. Next, choose the **OK** button; the dialog box will be closed and the member concentrated load will be added to the load group.

Applying Distributed Load

To assign a uniformly distributed load to a specific member along its full or partial span, select the member and choose the **Distributed** tool from the **Loads** panel of the **Member** tab; the **Add Member Distributed Load** dialog box will be displayed, refer to Figure 8-27. Select the load group from the **Load group** drop-down list and the load type from the **Load Type** drop-down list. Next, select the direction in which the load will act from the **Direction** drop-down list and enter the load values in the **Magnitude** edit box. Choose the **OK** button; the dialog box will be closed and the load will be applied to the selected members.

Applying Distributed Moment

To apply a uniformly distributed moment to a specific member along its full or partial span, select the member and choose the **Distributed Moment** tool from the **Loads** panel of the **Member** tab; the **Add Member Distributed Load** dialog box will be displayed, as shown in Figure 8-28. Next, select the load group from the **Load group** drop-down list and then moment type from the **Moment Type** drop-down list, respectively. Select the direction in which the load will act from the **Direction** drop-down list. Enter the load values in the **Magnitude** edit box and then choose the **OK** button; the dialog box will be closed and the load will be applied to the selected members.

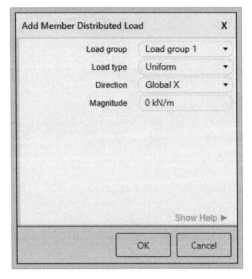

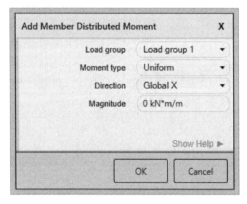

Figure 8-27 The ***Add Member Distributed Load*** *dialog box*

Figure 8-28 The ***Add Member Distributed Moment*** *dialog box*

Area Loads

The area load in STAAD.Pro Physical Modeller is used to distribute load to supporting members. The following section explains the methods to assign the area load.

Adding Area Load by Using Perimeter Nodes

To create an area load, select the perimeter nodes and then choose the **Area Load** tool from the **Create** panel of the **Model** tab. Next, click on nodes to form the perimeter of the area load and then double-click on the last node to stop adding nodes to the current area load.

Assigning Pressure on an Area Load

To assign uniform pressure on area loads, choose the **Assign Uniform Pressure & Direction** tool from the **Edit** panel of the **Area Loads** tab; the **Add area load** dialog box will be displayed, refer to Figure 8-29. Select the direction of load from the **Direction** drop-down list and enter the value of the magnitude in the **Magnitude** drop-down list. Next, choose the **OK** button; the dialog box will be closed and the uniform pressure will be applied.

Figure 8-29 The ***Add Area Load*** *dialog box*

Distributing Area Loads on Boundary Members

You can distribute the pressure applied on an area load to the bounding edge members. To do so, select the area load and then choose the **Distribute Pressure** tool from the **Edit** panel of the **Area Load** tab; the pressure load is distributed to a series of uniform and linearly varying loads along each bounding edge member. If you make changes to the area load pressure or bounding edges, you can repeat this procedure to update the loads distributed on members.

Example 3

You will create a portal frame with the same dimensions and section properties as given in Example 2. Next, you will assign supports and loads on the steel frame as described: Supports:- Node 1 to be Fixed support, Node 4 to be Pinned support. Load case 1 named as Load group 1, includes dead load and live load, Member 2 having a distributed load of 5 kN/m downward along global Y; Load case 2 named as Wind From Left which have 10KN point force at Node 2.

Steps required to complete this example are given below:

Step 1: Start Staad.Pro Connect Edition and then open Physical Model. Next, create the portal frame as discussed in Example 1. Assign the section and material properties to the model, as discussed in Example 2. Save the file by choosing the **Save As** option from the **File** menu and enter *c08_staad_connect_ex3* in the **File Name** edit box. Now, browse to the location *C:\STAAD Examples\c08_staad_connect* by clicking the button next to the **Location** edit box and choose the **Save** button.

Step 2: Next, select the lower-left node (N1). The node label appears if you have the labeling turned on. Choose the **Fixed** tool from the **Supports** panel of the **Node** tab. A fixed nodal support is assigned to this node. Select the lower-right node (N4). Choose the **Pinned** tool from the **Supports** panel of the **Node** tab; a pinned nodal support type is assigned to this node, as shown in Figure 8-30.

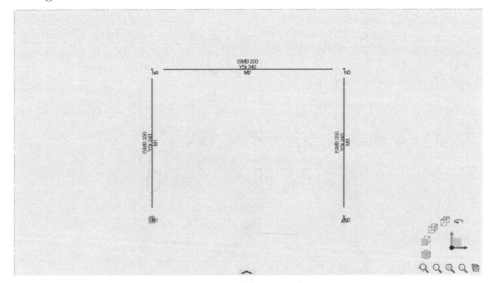

Figure 8-30 Portal frame with supports

Step 3: Select the entire structure. Choose the **Self Weight** button from the **Loads** panel of the **Model** tab; self weight multiplier with value -1 is assigned to the entire structure in the Y direction.

Step 4: Click the beam member (M2) to select it. Choose the **Distributed** tool from the **Loads** panel of the **Member** tab, the **Add Member Distributed Load** dialog box is displayed. Specify the load parameters, as shown in Figure 8-31. Choose the **OK** button; the load is applied to the beam, as shown in Figure 8-32.

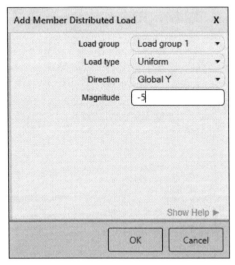

Figure 8-31 Add Member Distributed Load dialog box

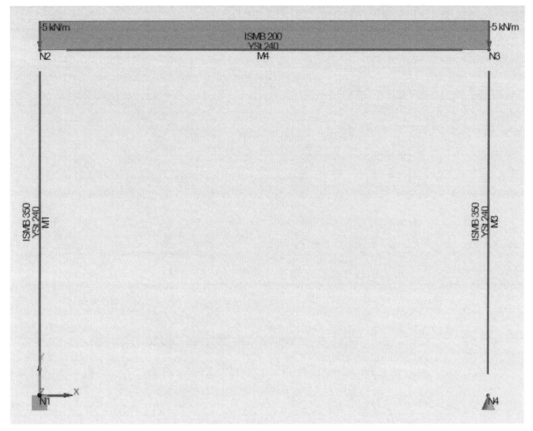

Figure 8-32 Distributed load on Portal Frame

Step 5: Choose the **Load Group** tool from the **Loads** panel of the **Model** tab; the **Add Load Group** dialog box is displayed. Specify the load group details, refer to Figure 8-33. Next, choose the **OK** button; the new load group is added to the **Loads > Load Group spreadsheet** and the load is automatically selected in the program of the window status bar as the current load group.

Note
When load items are created, they are added to the current load group.

Step 6: Select the upper-left node (N2). Choose the **Nodal Load** tool from the **Loads** panel of the **Node** tab; the **Add Nodal Load** dialog box is displayed. Specify the lateral wind load, refer to Figure 8-34. Next, choose the **OK** button.

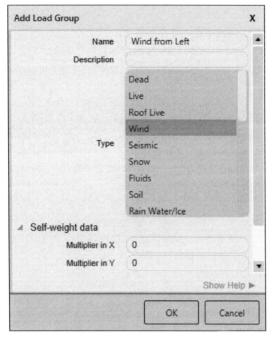

Figure 8-33 The **Add Load Group** dialog box

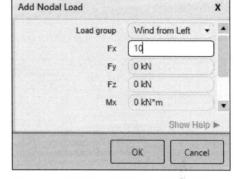

Figure 8-34 The **Add Nodal Load** dialog box

Step 7: The wind load is applied, as shown in Figure 8-35. Choose the **Save** option from the **File** menu to save the file. Then, close the file.

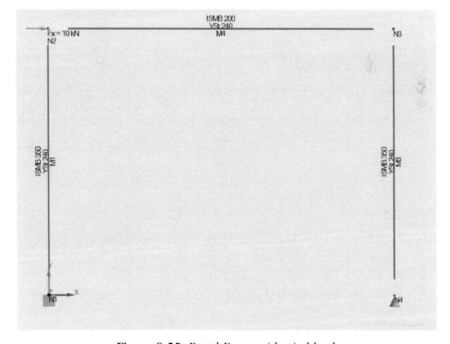

Figure 8-35 Portal Frame with wind load

STAAD.PRO MODEL DATA

STAAD.Pro Physical Modeller can create a physical model data and transfer it to STAAD. Pro Analytical Modeling for additional loading, analysis, and design. Any selections ade from design can also be updated in the physical model file. The following section explains how the physical model is transferred to an analytical model.

Transferring Physical Model to Analytical Model in STAAD.Pro

To transfer the STAAD.Pro physical model to analytical model, choose the **Create analysis model** option from the **File** tab; the **Return to Analytical Modeling** dialog box will be displayed. Refer to Figure 8-36 for the process of conversion of physical model to analytical model. Choose the **OK** button when the process of conversion is completed upto 100 percent; the physical model will be transferred to the analytical interface and the physical model interface will be closed.

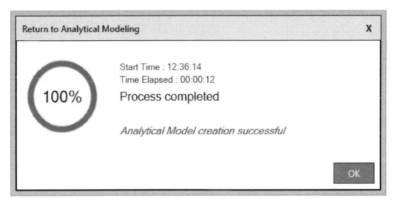

Figure 8-36 The **Return to Analytical Modeling** *dialog box*

Self-Evaluation Test

Answer the following questions and then compare them to those given at the end of this chapter:

1. The _____ tool is used to add beam between two nodes.

2. The _____ tool is used to assign material to the selected members.

3. Using the _____ tool, you can assign its own weight to \ structure.

4. The **Fixed** support is used to create roller, spring, and hinge supports. (T/F)

5. The **Fixed** supports are restrained in three translational degrees of freedom only. (T/F)

6. The **UCS** tool is used to add a Grid. (T/F)

Review Questions

Answer the following questions:

1. Which of the following tools is used to produce nodes in a straight line?

(a) **Member Generation**   (b) **Quadratic Generation**
(c) **Linear Generation**   (d) All of these

2. Which of the following tools is used to know the measurements of nodes and members?

(a) **Description**   (b) **Dimensions**
(c) **Sections**   (d) **Material**

3. Which of the following tools is used to add a point moment to the member?

(a) **Concentrated**   (b) **Distributed**
(c) **Concentrated Moment**   (d) **Distributed Moment**

4. Nodal Load is used to add concentrated load on the member. (T/F)

5. Custom support is used to add a roller load on the node. (T/F)

6. The **Load Group** tool is used to select a load group used in the structure. (T/F).

Answers to Self-Evaluation Test
1. Member, 2. Material, 3. Self Weight, 4. F, 5. T, 6. F

Index

Other Publications by CADCIM Technologies

The following is the list of some of the publications by CADCIM Technologies. Please visit *www.cadcim.com* for the complete listing.

Bentley STAAD.Pro Textbook
• Exploring Bentley STAAD.Pro V8i (SELECTseries 6)

Autodesk Revit Architecture Textbooks
• Exploring Autodesk Revit 2018 for Architecture, 14th Edition
• Exploring Autodesk Revit 2017 for Architecture, 13th Edition

Autodesk Revit Structure Textbooks
• Exploring Autodesk Revit 2018 for Structure, 8th Edition
• Exploring Autodesk Revit 2017 for Structure, 7th Edition

Autodesk Revit MEP Textbooks
• Exploring Autodesk Revit 2018 for MEP, 5th Edition
• Exploring Autodesk Revit 2017 for MEP, 4th Edition

AutoCAD Civil 3D Textbooks
• Exploring AutoCAD Civil 3D 2018, 8th Edition
• Exploring AutoCAD Civil 3D 2017, 7th Edition

AutoCAD Map 3D Textbooks
• Exploring AutoCAD Map 3D 2018, 8th Edition
• Exploring AutoCAD Map 3D 2017, 7th Edition

RISA-3D Textbook
• Exploring RISA-3D 14.0

Autodesk Navisworks Textbooks
• Exploring Autodesk Navisworks 2017, 4th Edition
• Exploring Autodesk Navisworks 2016, 3rd Edition

AutoCAD Raster Design Textbooks
• Exploring AutoCAD Raster Design 2017
• Exploring AutoCAD Raster Design 2016

AutoCAD Textbooks
• AutoCAD 2018: A Problem-Solving Approach, Basic and Intermediate, 24th Edition
• Advanced AutoCAD 2018: A Problem-Solving Approach (3D and Advanced), 24th Edition
• AutoCAD 2017: A Problem-Solving Approach, Basic and Intermediate, 23rd Edition
• AutoCAD 2017: A Problem-Solving Approach, 3D and Advanced, 23rd Edition

Autodesk Inventor Textbooks
• Autodesk Inventor Professional 2018 for Designers, 18th Edition
• Autodesk Inventor Professional 2017 for Designers, 17th Edition

AutoCAD MEP Textbooks
• AutoCAD MEP 2018 for Designers, 4th Edition
• AutoCAD MEP 2016 for Designers, 3rd Edition

Solid Edge Textbooks
• Solid Edge ST10 for Designers, 15th Edition
• Solid Edge ST9 for Designers, 14th Edition

NX Textbooks
• NX 12.0 for Designers, 11th Edition
• NX 11.0 for Designers, 10th Edition

Mold Design Using NX Textbook
• Mold Design Using NX 11.0: A Tutorial Approach

NX Nastran Textbook
• NX Nastran 9.0 for Designers

SolidWorks Textbooks
• SOLIDWORKS 2018 for Designers, 16th Edition
• SolidWorks 2014: A Tutorial Approach
• Learning SolidWorks 2011: A Project-Based Approach

SolidWorks Simulation Textbooks
• SOLIDWORKS Simulation 2018: A Tutorial Approach
• SOLIDWORKS Simulation 2016: A Tutorial Approach

CATIA Textbooks
• CATIA V5-6R2017 for Designers, 15th Edition
• CATIA V5-6R2016 for Designers, 14th Edition

Creo Parametric and Pro/ENGINEER Textbooks
• Creo Parametric 4.0 for Designers, 4th Edition
• PTC Creo Parametric 3.0 for Designers, 3rd Edition
• Pro/Engineer Wildfire 5.0 for Designers

AutoCAD Plant 3D Textbooks
• AutoCAD Plant 3D 2018 for Designers, 4th Edition
• AutoCAD Plant 3D 2016 for Designers, 3rd Edition

ANSYS Textbooks
- ANSYS Workbench 14.0: A Tutorial Approach
- ANSYS 11.0 for Designers

Creo Direct Textbook
- Creo Direct 2.0 and Beyond for Designers

Autodesk Alias Textbooks
- Learning Autodesk Alias Design 2016, 5th Edition
- AliasStudio 2009 for Designers

AutoCAD LT Textbooks
- AutoCAD LT 2017 for Designers, 12th Edition
- AutoCAD LT 2016 for Designers, 11th Edition

EdgeCAM Textbooks
- EdgeCAM 11.0 for Manufacturers
- EdgeCAM 10.0 for Manufacturers

AutoCAD Electrical Textbooks
- AutoCAD Electrical 2018 for Electrical Control Designers, 9th Edition
- AutoCAD Electrical 2017 for Electrical Control Designers, 8th Edition

Autodesk 3ds Max Design Textbooks
- Autodesk 3ds Max Design 2015: A Tutorial Approach, 15th Edition
- Autodesk 3ds Max Design 2014: A Tutorial Approach

Autodesk 3ds Max Textbooks
- Autodesk 3ds Max 2018: A Comprehensive Guide, 18th Edition
- Autodesk 3ds Max 2018 for Beginners: A Tutorial Approach, 18th Edition

Autodesk Maya Textbooks
- Autodesk Maya 2018: A Comprehensive Guide, 10th Edition
- Character Animation: A Tutorial Approach

Pixologic ZBrush Textbooks
- Pixologic ZBrush 4R7: A Comprehensive Guide, 3rd Edition
- Pixologic ZBrush 4R6: A Comprehensive Guide

Fusion Textbooks
- Blackmagic Design Fusion 7 Studio: A Tutorial Approach
- The eyeon Fusion 6.3: A Tutorial Approach

Flash Textbooks
- Adobe Flash Professional CC 2015: A Tutorial Approach
- Adobe Flash Professional CC: A Tutorial Approach

Computer Programming Textbooks
• Introducing PHP/MySQL
• Introduction to C++ programming, 2nd Edition
• Learning Oracle 12c - A PL/SQL Approach
• Learning ASP.NET AJAX
• Introduction to Java Programming, 2nd Edition
• Learning Visual Basic.NET 2008

MAXON CINEMA 4D Textbook
• MAXON CINEMA 4D R19 Studio: A Tutorial Approach, 5th Edition

Oracle Primavera Textbook
• Exploring Oracle Primavera P6 R8.4

AutoCAD Textbooks Authored by Prof. Sham Tickoo and Published by Autodesk Press
• AutoCAD: A Problem-Solving Approach: 2013 and Beyond
• AutoCAD 2012: A Problem-Solving Approach
• AutoCAD 2011: A Problem-Solving Approach
• AutoCAD 2010: A Problem-Solving Approach
• Customizing AutoCAD 2010
• AutoCAD 2009: A Problem-Solving Approach

Coming Soon from CADCIM Technologies
• SolidCAM 2016: A Tutorial Approach
• Autodesk Fusion 360: A Tutorial Approach
• Project Management Using Microsoft Project 2016 for Project Managers

---

Online Training Program Offered by CADCIM Technologies
CADCIM Technologies provides effective and affordable virtual online training on animation, architecture, and GIS softwares, computer programming languages, and Computer Aided Design, Manufacturing, and Engineering (CAD/CAM/CAE) software packages. The training will be delivered 'live' via Internet at any time, any place, and at any pace to individuals, students of colleges, universities, and CAD/CAM/CAE training centers. For more information, please visit the following link: *http://www.cadcim.com*.

Made in the USA
Middletown, DE
27 December 2019